ANIMAL DOMESTICATION

Edited by **Fabrice Teletchea**

Animal Domestication
http://dx.doi.org/10.5772/67910
Edited by Fabrice Teletchea

Contributors

Fabrice Teletchea, Pascal Fontaine, Alain Pasquet, Joao L Saraiva, Maria F. Castanheira, Pablo Arechavala-López, Jenny Volstorf, Billo Heinzpeter Studer, Thomas Lecocq, Mirte Bosse

First published in London, United Kingdom, 2019 by IntechOpen
IntechOpen is the global imprint of INTECHOPEN LIMITED, registered in England and Wales, registration number: 11086078, The Shard, 25th floor, 32 London Bridge Street
London, SE19SG – United Kingdom
Printed in Croatia

British Library Cataloguing-in-Publication Data
A catalogue record for this book is available from the British Library

Additional hard and PDF copies can be obtained from orders@intechopen.com

Animal Domestication, Edited by Fabrice Teletchea
p. cm.
Print ISBN 978-1-83880-174-8
Online ISBN 978-1-83881-133-4
eBook (PDF) ISBN 978-1-83881-134-1

Meet the editor

Fabrice Teletchea is an associate professor at the Université of Lorraine (France). He received a master's degree from the National Museum of Natural History in Paris, France, and a PhD in Systematics from the Université of Lyon in 2005, after which he studied fish domestication in aquaculture. Dr. Teletchea developed a comparative framework of the reproductive strategies of European freshwater fish species based on a new database called STOREFISH, in order to better understand the different trade-offs observed at the early life stages of fish and to help domesticate fish more efficiently. In parallel, he teaches ecology, fish biology, animal domestication, and managed a diploma entitled "Inland Aquaculture and Aquariology" at the IUT Nancy-Brabois.

Contents

Preface

Domestication, which is by definition a long ongoing process, was one of the most significant cultural and evolutionary transitions of human history. Domestication was indeed a core component of a major change in the way of life of an increasing number of human societies throughout the world, in a process called Neolithisation. This process constitutes also a fundamental change in the evolution of the biosphere, mainly due to the development of agriculture, which is now responsible for the transformation of approximately 40 percent of the earth's surface.

Ever since Darwin, the study of domestication has puzzled scientists. Each year, hundreds of articles, as well as several books, are published in numerous disciplines, among which include archaeology, ethnoarchaeology, genetics, evolutionary biology, and zootechny. Nevertheless, despite this vivid interest, both the terms "domestication" and "domestic animal" remain confusing and poorly defined, and several animal groups are still poorly studied, such as fish and insects.

Because there is no scientific reason to consider the domestication of land and aquatic animals differently, this book groups scholars working both on different topics and animal taxa. Fruitful exchanges between these diverse scholars could bring new insights to both the concept of domestication itself and the differences between wild and domesticated animals. On a more applied view, this could also help to better domesticate species in the future as farmed animals are continuously evolving generations after generations, particularly in response to changes in technology and husbandry practices, which are also evolving and constantly improving, and to global change.

The book includes seven chapters, three on land animals and four on aquatic animals. The first chapter presents a brief overview of the domestication of land animals, focusing on the five major farmed species (cattle, pig, sheep, goat, and horse). The second chapter provides an in-depth overview of the complex process of domestication, admixture, and selection leading towards the genetic diversity in extant animal breeds, using pig as the model. The third chapter presents an overview of the main ancient and recent insect domestication histories and rereads them b through the lens of the domestication process, pathways, triggers, and consequences observed in other animal species. The fourth chapter briefly assesses the domestication of fish, by focusing on seven of the main farmed species globally (common carp, Nile tilapia, rainbow trout, striped catfish, Atlantic salmon, European seabass, and Atlantic Bluefin tuna). The fifth chapter reviews behavioural traits in hatchery-reared fishes that have often been altered in a characteristic manner by domestication. The sixth chapter is an extensive discussion of the welfare of fish over the course of domestication. Finally, the seventh chapter provides a comprehensive analysis of the domestication history of Eurasian perch in Europe.

It is my hope that this book will stimulate new discussions among scholars working on different topics and animal groups.

Fabrice Teletchea
UR AFPA
Université de Lorraine
Vandoeuvre-Les-Nancy, France

Chapter 1

Animal Domestication: A Brief Overview

Fabrice Teletchea

Additional information is available at the end of the chapter

http://dx.doi.org/10.5772/intechopen.86783

Abstract

In the origin of modern humans, hunting of wild animals and gathering of wild plants in nature were the primary subsistence strategies. Yet, about 12,000 years ago, the domestication of plants and animals began. The two main goals of the present chapter are to briefly describe (i) how wild animals were domesticated and (ii) what are the main biological consequences for the major farmed species (cattle, pig, sheep, goat, and horse). During about 98% of their domestication history, domestic animals have been managed in a sustainable way by farmers, followed by a period of strong selection about 200 years ago to produce hundreds of well-defined breeds. A few decades ago, the selection pressures have further increased, leading to a few industrial breeds, which were introduced in numerous countries, most often at the expense of local breeds. Within a few decades, we thus might lose most of the highly valuable farm animal genetic resources that humans have gradually selected over the past millennia. Consequently, priorities should be given to preserve the genetic resources in marginal or rare breeds, and selection programs should aim at restoring the genetic diversity in industrial breeds.

Keywords: domestication, domesticated animals, genetics, industrial breed, local breed

1. Introduction

Since the origin of modern humans, *Homo sapiens*, about 200,000 years ago, hunting of wild animals and gathering of wild plants in nature were the primary subsistence strategies [1]. Yet, about 12,000 years ago, at the end of the most recent ice age and during the transition to the present interglacial period, domestication of plants and animals began [1–5]. This phenomenon occurred in at most nine areas of the world: the Fertile Crescent, China, Mesoamerica, Andes/Amazonia, Eastern United States, Sahel, tropical West Africa, Ethiopia, and New Guinea [4]. From these handful homelands of agriculture, a restricted number of domesticated species were progressively introduced across the globe as farmers migrate to new regions [4].

This is because the control of food production conferred to farmers huge demographic, technological, political, and military advantages over neighboring hunter-gatherers, allowing them to impose their lifestyle [4]. The history of the past millennia consists of tales of hunter-gatherer societies becoming driven out, infected, conquered, or exterminated by farming societies in all areas across the world suitable for farming [3, 4].

The domestication of plants and animals was part of a major transformation in the way of life of an increasing number of human societies, with deep social and spiritual changes, called the Neolithic transition [5, 6]. This also enabled a strong increase of the human population from about 1 million during the millennia before the advent of agriculture [7] to more than 7 billion today [8]. The increase of human population is spectacular during the past decades (**Figure 1**), with an additional 4 billion people since 1960 [8].

In 2010, the world agricultural production reaches more than 7.6 billion tons [7], representing a three-fold increase compared to 1961 (**Figure 2**). Globally, less than two-thirds of crop production (on a mass basis) are allocated to human food, versus 35% to animal feed, and 3% for bioenergy or other industrial products [9]. It is, however, important to highlight that global food production relied *in fine* on a tiny fraction of wild species domesticated in the past millennia, representing about 0.08% of known land plant species and 0.0002% of known land animal species [10]. Only about 15 plant species and less than 10 animal species supply more than 90% of worldwide agriculture production [11]. Four crops (wheat, rice, corn, and potato) account for more food production than all other crops combined [11]. Inversely, hunting and gathering have today become secondary (and most often recreational) activities that contribute little to global food security [12], one significant exception being the consumption of wild meat in a few regions, notably in Central Africa [13]. This implies that even tough humans consumed diverse food products across the globe; they mostly come from the same domesticated plant and animal species. In the past decades, the standardization of food products has also strongly increased with the spread of few multinational food companies, such as McDonald's or Subway.

The other main consequences of domestication are that the bulk of global agriculture is today based on the culture or farming of a few alien domesticated species that had been

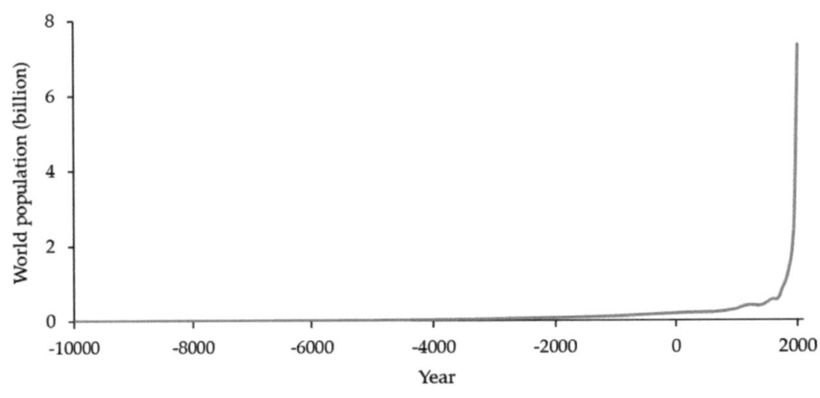

Figure 1. World population growth over time (modified from [8]).

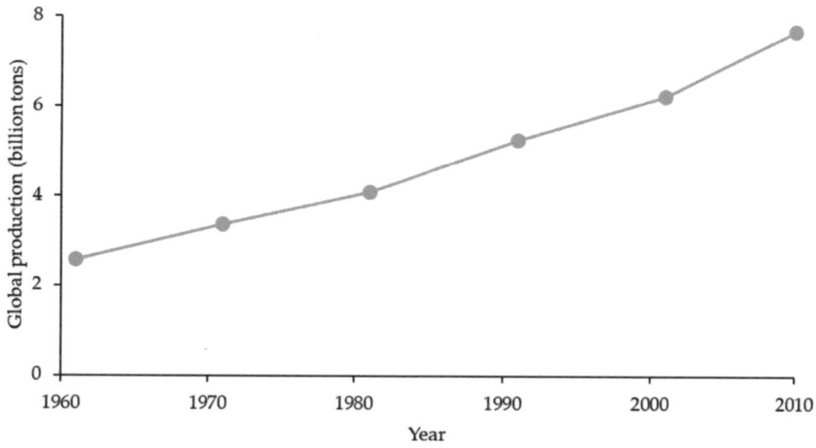

Figure 2. Global agricultural production, 1961–2010 (modified from [7]).

progressively introduced in all continents. This has contributed to widespread faunal and floral homogenization [14]. Nevertheless, because alien species are present for a very long period, they are generally not perceived as exogenous or introduced [15], but rather as part of the natural landscape [16–18]. This phenomenon has been described as the shifting baseline syndrome [19]. Agriculture is today responsible for the destruction or modification of nearly 40% of the land surface [20]. For instance, about 7 to 11 million km^2 of forest have been lost in the past 300 years due to land-use activities, primarily for agricultural expansion and timber extraction [20]. Besides, intensification of agriculture has also resulted in the degradation of water quality in numerous freshwater and coastal ecosystems due to the global use of fertilizers, pesticides, and antibiotics [9, 20]. Modern agriculture is thus generally considered to be the primary destructive force of biodiversity [17], which has led to the sixth mass extinction [21]. Some scientists even consider that truly wild nature (pristine zones from human impacts) does no longer exist on Earth [22]. In 2002, Crutzen [23] proposed to assign the term "Anthropocene" to the present geological epoch, supplementing the Holocene, once humans have become an important geochemical force and perhaps the dominant ecological force on the planet. The Anthropocene era could be said to have started in the late eighteenth century [23]. In conclusion, domestication corresponds to a pivotal change in the history not only of humanity but also of the biosphere [5, 6].

The two main goals of the present chapter are to briefly describe (i) how wild animals were domesticated and (ii) what are the main biological consequences for the major farmed species.

2. How were animals domesticated?

Domestication is a long and endless process by which animals become adapted to both humans and captive conditions ([24–26]; for an overview of definitions of domestication, see

[27]). Three main pathways of domestication have been proposed for land animals: a commensal pathway, a prey pathway, and a directed pathway [6, 28–30]. In the commensal pathway, the animals themselves played the largest role [29]. The animals first move into an anthropogenic habitat, most likely spurred by an attraction to human waste, and later develop a two-way partnership with humans [29]. Several domesticated species have followed this path, among which are dog (*Canis familiaris*), cat (*Felis catus*), or chicken (*Gallus domesticus*) [29]. In the prey pathway, humans have initiated domestication, perhaps as a response to depletion of local stocks of prey animals that humans had hunted for thousands of years [31], to enhance the yield or predictability of a resource (meat or hides) [29]. Over time and under certain circumstances, these game management strategies developed into actual herd management and, eventually, the controlled breeding of managed animals [28]. The main species that followed this pathway are sheep (*Ovis aries*), goat (*Capra hircus*), or cattle (*Bos taurus*) [29]. In the direct pathway, humans deliberately set out to domesticate a species [28, 31]. This pathway skips the early phases of habituation and management and starts with the capture of wild animals with the deliberate intention of controlling their reproduction [29]. This pathway occurred more rapidly and was accompanied by a dramatic bottleneck [29]. The main species are horse (*Equus caballus*), donkey (*Equus asinus*), and dromedary (*Camelus dromedarius*) [28].

Species that followed either commensal or prey pathways tend to possess more traits that make them appropriate candidates for domestication. Conversely, species on directed pathways likely possess barriers to domestication that require more knowledge on the part of humans to overcome [28, 31].

Whatever the pathway followed, captive animals began to be domesticated at some point. Yet, as for domestication, there is no consensus today about what a domesticated species is (see [27] for a review of the main definitions). Nevertheless, most authors considered that a domesticated species is a group of animals reproduced in captivity and modified from their wild congeners [27]. Yet, wild and domesticated animals should not be considered as complementaries (such as true/false, dead/alive) but rather as antonyms (such as long/short, fast/slow) because they represent the extremes of a process and not a simple dichotomy [32]. In other words, there is not a clear biological separation between wild and domesticated animals [33]. In addition, a domesticated animal is neither in a final nor a static status, and thus farmed species are still evolving today, particularly in response to changes in technology and husbandry practices, which themselves are evolving and constantly improving [34]. Conversely, domesticated species can sometimes return to nature, a process known as feralization [35].

3. How have animals evolved during domestication?

During domestication, five main genetic processes were involved [15, 28, 34], including inbreeding and genetic drift (two uncontrolled processes), natural selection in captivity and relaxation of natural selection (two partially controlled processes), and active selection (one controlled process) [34, 35]. The two uncontrolled processes are due to the limited size of the population (known as inbreeding) and the random changes in gene frequencies (genetic drift).

The two partially controlled processes are natural selection in captivity that accounts for selection imposed on captive populations that cannot be attributed to active (or artificial) selection and relaxation of natural selection expectably accompanying the transition from wild to captive environments [35]. At last, the fifth genetic process is controlled, known as active selection, because changes are directional [34, 35].

Domesticated animals have been profoundly modified during domestication. Indeed, the variation range of certain traits within a domesticated species occasionally exceeds that in whole families or even orders [36, 37]. Modifications resulting from domestication concern morphoanatomy, physiology, behavior, and genetics [31, 35, 38–40]. Behavior is probably the first to have been modified during domestication [35]. Nevertheless, behavioral traits neither appeared nor disappeared during domestication but rather are the response thresholds that changed [34, 35]. One of the most remarkable behavioral changes shared by all domesticates is their tolerance of proximity to (or complete lack of fear of) people [31, 37, 39]. Besides, because humans provide shelter, food, and protection against predators, domesticated animals most often express a lower incidence of antipredator behaviors and show lower motivation for foraging [34]. More generally, mood, emotion, agnostic and affiliative behavior, as well as social communication all have been modified in some way by domestication [39]. Most domesticated animals are also more precocious than their wild counterparts [34]. The activity of their reproductive system became enhanced and relatively uncoupled from the environmental photoperiod, and they all acquired the capacity to reproduce in any season and more often than once a year [37]. At last, the most spectacular and obvious changes concern morphology, among which are the animal size (dwarfs and giants), proportions (fewer vertebrae, shorter tails), color, length and texture of coat, wavy or curly hair, rolled tails, and floppy ears or other manifestations of neoteny (the retention of juvenile features into sexual maturity) [37, 39]. In most domesticated species, head or brain size has decreased [34]. The most illustrative example of such considerable changes is the morphological variations in dogs [37]. These morphological changes ("domestication syndrome") may all be linked to strong selection for lowered reactivity to external stimuli [31]. At the beginning of the twentieth century, modern breeding programs were initiated, leading to dramatic changes in productivity, e.g., increase laying rate for laying hens or improved feed conservation ratio, meat yield, and growth rate in broiler chickens [41].

4. A brief history of the major domesticated animals

Even though the decision to consider farmed or captive animals as domesticated is subjective and arbitrary [35, 41], most authors agree that about 40 species around the world that directly or indirectly contribute to agriculture are domesticated; this number varies between 20 and 50 following the definitions used for a domesticated animal [36, 42–44]. Several of those domesticated species have a distinct scientific name than their wild ancestors [25].

The 14 most important domesticated mammal species are indicated in **Table 1**, among which the domestication of the "big five" (cattle, pig, sheep, goat, and horse) [3, 4] are further

Common name	Scientific names	Partial list of potential wild progenitors (in bold the main one)	Approximate date of domestication: BP	Number of breeds	Pathway to domestication
Sheep	*Ovis aries*	***O. orientalis, O. musimon***	9000	850	Prey pathway
Goat	*Capra hircus*	***C. aegagrus***, *C. falconeri*	9000	320	Prey pathway
Cow, cattle	*Bos taurus* and *B. indicus*	***B. primigenius, B. namadicus***	8000	815	Prey pathway
Pig	*Sus domesticus*	***S. scrofa***, *S. celebensis, S. barbatus*	8000	350	Commensal pathway
Horse	*Equus caballus*	***E. ferus***, *E. przewalski*	6000	350	Directed pathway
Dromedary	*Camelus dromedarius*	***C. dromedarius***	4500	50	Directed pathway
Bactrian camel	*Camelus bactrianus*	***C. ferus***, *C. bactrianus*	4500	6	Directed pathway
Llama and alpaca	*Lama glama* and *L. pacos*	***L. guanicoe*** and *V. vicugna* (?)	6000	2 + 2	Prey pathway
Donkey or ass	*Equus asinus*	***E. africanus***	6000	70	Directed pathway
Reindeer	*Rangifer tarandus*	***R. tarandus***			
Water buffalo	*Bubalus bubalis*	***B. bubalis***	6000	70	Prey pathway
Yak	*Bos grunniens*	***B. grunniens***	4500		Prey pathway
Bali cattle	*Bos javanicus*	***B. javanicus***			Prey pathway (?)
Mithan	*Bos frontalis*	***B. frontalis***			Prey pathway (?)

BP, before present. If no information was found, cells were left empty. Note that the number of breeds per species varies between authors.

Table 1. List of the world's 14 valuable big domestic mammals, including the major 5 (in bold) followed by the minor 9 [3, 4, 25, 28, 42, 44].

described below. For the five most valuable species, the domestication resulted in the creation of hundreds of breeds, particularly in the past centuries [42, 45, 46]. In France, the article D.653.9 of the rural code defines breed for ruminant species as "a group of animals that share sufficient common features to be considered homogeneous by one or several groups of breeders that agree on the broodstock renewal and induced changes, including the international level" [47]. Breeds have therefore both a biological sense (common features) and a social acceptance (group of breeders); the relative importance of the latter increased in the past years, for scientists as well as in the application of policies [47].

4.1. Cattle

The wild ancestor of cattle is a group of races of the now extinct aurochs *Bos primigenius* [48–50]. The aurochs, the last specimen of which died in a Polish park in 1627, had a very wide geographic distribution, which extended from East Asia to Europe and North Africa [42, 48, 50].

Traditionally, two major types of domestic cattle are considered: zebu (*Bos indicus*) which have a prominent thoracic hump and taurine (*Bos taurus*), which do not [40, 42, 49, 50]. However, these two species fully interbreed, and a meta-analysis of different microsatellite datasets revealed taurine-zebu admixture over Europe, southwest Asia, and Africa [40, 45, 49, 50]. Molecular evidence suggest that these two species came from two independent domestication events: zebu cattle were domesticated in the Indus valley region ca. 8000–7500 B.P., whereas taurine cattle were domesticated in Anatolia 10,500–10,000 B.P. [40, 42, 50–52]. However, Larson and Burger [29] recently suggested that only the latter was domesticated, while zebu may have resulted from the introgression of wild zebu populations into taurine cattle that were transported eastward. During several millennia, extensive gene flow among different groups of domestic cattle, as well as with aurochs until its extinction, was possible, leading to relatively high effective population sizes and preventing genetic drift at the regional scale [40, 48, 50, 51]. This might partly explain the relatively large cattle gene pool despite a likely bottleneck at the time of domestication [50]. Besides, it is also possible that other species were crossed with cattle in some areas of the world, including the yak (*Bos grunniens*) in Nepal or banteng (*Bos javanicus*) in Southeast Asia and Indonesia, which also contribute to maintain or increase genetic variability [40]. The large size of cattle and its low growth, as well as the early use for milk or traction, imply relatively low levels of directed selection during millennia [51]. However, this situation changed dramatically about 200 years ago with the emergence of breed concept [50]. The first cattle herd book was published in Britain in 1822 [49]. Since that time, stronger selection pressures have been applied to local populations followed by standardization of the desired conformation and performance, such as high milk yield for dairy cattle breeding programs [49]. This led to an isolation of breeds from each other (ca. 800 are now recognized; see **Table 2**), which could have caused a genetic drift and inbreeding and perhaps a fitness decrease [40, 46, 50]. Nevertheless, gene flow between neighboring regions did not completely stop, as deliberate upgrading was realized in order to increase production characteristics by using bulls of other populations from the same or a different country [45]. More recently, the number of males involved in reproduction schemes has drastically decreased with the expansion of artificial insemination, leading to another strong reduction of effective population size of breeds and inexorably to a genetic drift and loss of alleles [46, 50, 63]. For example, at the worldwide level, the Holstein cattle has an effective population size of about 50 [50]. This strong decrease of the effective population size might explain the strong reduction in fertility as well as the genetic diseases observed in this breed [50]. An even more extreme result was found in Japan, where the Japanese black cattle had an effective population size of 17.2 in between 1993 and 1997, despite a census size of 0.53 million reproductive cows [49]. Another extreme case of low genetic variability is a feral British breed, Chillingham cattle, for which 24 out of 25 microsatellite loci were found homozygous [46]. Inversely, numerous cattle breeds still have substantial nucleotide diversity, indicating a large ancestral effective population size [46]. In the past decades, a few of the most productive breeds were imported throughout the world at the expense of local, apparently less productive populations [45].

4.2. Pig

The wild ancestor of domestic pigs is boar *Sus scrofa* [42, 64]. Wild boars occurred throughout Eurasia and North Africa [42]. Multiple independent domestication events, mainly in Asia

Minor, Europe, and East Asia, have probably occurred, starting approximately 9000 years ago [40, 52, 64–66]; the earliest remains of domesticated pigs have been excavated at Çayönü in Southeast Anatolia [65]. Chinese breeds originated in East Asia, whereas European breeds are believed to have originated in Southwest Asia [42, 67]. These domestication events were separated not only by thousands of kilometers but also by thousands of years [65]. During millennia, it is likely that out-crossing of domestic pigs with wild boar was common in traditional pig husbandry across Europe [52, 67]. By the late middle ages, European and Asian domestic pigs were genetically very different because they were based on wild boar populations that diverged around 1 million years ago, and for thousands of years, they were submitted to selection pressures on very different traits [64, 68]. By the late eighteenth to early nineteenth century, strict organized breeding was adopted to improve and develop livestock breeds, particularly in Britain, as a reaction to increasing demand for meat in the wake of the industrial revolution [64, 66]. European breeders turned also their attention to Asia and imported Chinese pigs to improve their breeding stock [40, 64–66]. From the eighteenth century, pig breeds were selectively bred for specific production traits such as early maturation, rapid growth, and increased prolificacy. In addition, the coat color phenotype (which includes both skin and hair pigmentation) was another morphological trait often used during the selective breeding process. Substantial changes (body size, color, body shape, skull morphology, ear carriage, behavior, prolificacy, teat number, and other traits) occurred in breeds over a short period of time, resulting in the development of numerous distinct pig breed phenotypes [64–66]. From the twentieth century, with the recognition of the benefits of genetic improvement and changing consumer preferences, certain pig breeds experienced further strong selection for lean meat content, muscularity, and enhanced reproduction [64, 66]. To date, there are likely over 730 pig breeds or lines globally of which two thirds are in China and Europe and over 270 are considered as endangered or critical. Currently, 58 pig breeds are recorded as "transboundary" (occurring in more than one country) including 25 regional transboundary breeds and 33 international transboundary breeds [65]. The worldwide distribution of pigs is dominated by five international transboundary pig breeds from the United States (USA) or Europe, i.e., Large white (117 countries), Duroc (93 countries), Landrace (91 countries), Hampshire (54 countries), and Pietrain (35 countries) [65].

4.3. Sheep

The wild ancestors of the domestic sheep are probably the mouflon (*Ovis musimon*) and the urial (*Ovis orientalis*) [42, 49, 50]. Both archaeological and genetic data spot the domestication center of sheep in eastern Anatolia and North-West Iran [50] between 8500 [49] and 12,000 years ago [40]. The sheep mitochondrial DNA polymorphism diversity and single-nucleotide polymorphism (SNP) diversity seem to support an absence of a genetic bottleneck, and thus domestication occurred from a broad genetic base [50, 69]. Sheep were first farmed for access to meat before human-mediated specialization for wool and milk commenced ca. 4000–5000 years ago [69]. It has recently been shown that particular regions of the genome contain strong evidence for accelerated change in response to artificial selection, such as the removal of horns, likely to be one of the oldest morphological modifications that accompanied domestication and a trait now common across many modern breeds [69]. Furthermore, other genomic regions under selection in sheep contain genes controlling pigmentation, reproduction, and body size [69]. In the last few hundred years, the division of animals into breeds, followed by the identification of superior rams and their disproportionate genetic contribution via artificial insemination, has lifted the pace of genetic gain for production traits [69]. Extensive haplotype sharing and generally low divergence time between breeds reveal that frequent genetic exchange has occurred during the development of modern breeds [69]. Approximately 75% of modern sheep breeds have retained an effective population size in excess of 300 [69]. The number of breeds is comprised between 850 and 1409 [49]. Yet, many sheep breeds originally selected for good performance in a specific, sometimes isolated, geographical area (e.g., the Shetland, Soay, or Herdwick breeds) are now considered rare. With generalist-type sheep taking over the larger part of intensive sheep production, maintaining genetic diversity by conserving these traditional breeds has become a challenge [50, 70].

4.4. Goat

The wild ancestor of goat is the bezoar, *Capra aegagrus* [40, 42, 49]. The first archaeological evidence of goat domestication traces back in the Fertile Crescent about 10,000 years ago [42, 49, 71]. A large-scale analysis of current bezoar mitochondrial DNA (mtDNA) polymorphism over its whole geographic distribution suggested that the domestication process occurred over a very large area encompassing eastern Anatolia and North-West Iran [50]. Additional primary centers of goat domestication, including the Indus Valley, Southern Levant, and China, have not been convincingly demonstrated yet [71]. Analysis of the goat mitochondrial DNA polymorphism of the main haplogroup (representing more than 90% of the haplotypes) strongly supports the absence of bottleneck at the domestication time in goats [50]. Besides, goat mtDNA polymorphism also suggests high historical gene flow among continents, which already occurred during the Neolithic expansion into Europe [49]. The extraordinary adaptability and hardiness of goats favored their rapid spread over the Old World [71]. Goats have successfully adapted to desert, mountainous, and tropical areas where other livestock species would not thrive [71]. Between the fifteenth and eighteenth centuries, goats were transported to America and Oceania [71]. Over the course of domestication, several morphological traits were modified, such as horn and ear shapes, the presence of wattles, long hair, and coat colors, which were driven probably by intentional selection as well as by genetic drift, isolation, and founder effects [71]. Throughout the ages, goats have been raised for milk production and cheese, meat, and skin and fiber commodities such as leather, mohair wool, and cashmere hair [71]. Breeds also show strong differences in their physiological capacity of adaptation to extreme conditions of temperature and humidity and differ in feed efficiency, behavior, and resistance to infectious and parasitic diseases [71]. Today, Asia and Africa contain 58.2 and 36.2%, respectively, of the 1 billion goats worldwide with much smaller populations in Europe (1.7%), America (3.5%), and Oceania (0.4%) [71]. China (187.8 million heads), India (133 million heads), Nigeria (71 million heads), Pakistan (66.6 million heads), and Bangladesh (55.9 million heads) are the top five goat producers [71]. In most countries, the sustained growth of the world goat population during the last 50 years (from 368 million heads in 1964 to 1006 million heads in 2014) has not involved a general improvement of the production and reproduction techniques associated with their management. Well-organized selection programs are in contrast restricted to a few highly productive dairy breeds from Europe, North America, and Australia [71]. Despite its low caprine census, Europe produces 17.5 and 42.5% of the goat milk and cheese, respectively, consumed around the world, reflecting the benefits of raising highly selected dairy breeds such as the Saanen, Alpine, and Toggenburg under semi-intensive or intensive conditions [71]. The main traits under selection are milk production, protein and fat contents, somatic cell count, and udder morphology [71]. Globally, it has been proposed that 500–600 goat breeds exist; yet it is difficult to establish a reliable figure because several local populations are not managed via phenotypic standardization, herd book registration, and controlled reproduction [71]. Among these breeds, 86 have spread beyond their countries of origin and are considered cosmopolitan or transboundary breeds, whereas 19 have become extinct, and about 90 are critically endangered or just endangered (such as Arapawa, Bagot, Golden Guernsey, San Clemente, and Mallorquina), whereas 157 are not at risk [71]. Population decline of local goat breeds is mostly due to their replacement or uncontrolled crossbreeding with more productive foreign varieties, the progressive abandonment of low income rural activities, and the lack of genetic conservation programs [50, 71]. Some goat populations have escaped from captivity and became feral [71].

4.5. Horse

The wild ancestor of domestic horse is the now extinct, *Equus ferus* from central Asia [52]. The Asian wild horse, *Equus przewalskii* [42], also significantly contributed to the genetic makeup of domestic horses [72]. Even though there have been no confirmed sightings of wild Przewalski's horses since 1966, the species has been maintained in captivity for the last 90 years [42]. In addition to Przewalski's horse, a third divergent lineage corresponding to a wild population that inhabited the Holarctic region has also contributed to the genome of modern domestic horses [72]. Both archaeological and genetic evidence strongly support the onset of domestication of horse in the western Eurasian Steppes of Ukraine dating to 5500 years ago [42, 72]. Over the course of domestication, it has been argued that difficulties in maintaining domestic horse herd sizes during pastoral migrations led directly to restocking through the capture of wild females [52, 72]. Horses were not only used as a source of meat and milk; their stamina and quickness provide humans with rapid transportation, which has considerably changed the speed and magnitude of the circulation of goods and people, as well as cultural exchange, including the spread of Indo-European languages, religions, science, and art, and diseases [72]. With the introduction of the horse collar and horseshoes in agriculture, the horse was increasingly used for tilling soils, incrementing farmland productivity in medieval Europe, and remains today a crucial asset to the agriculture of the least-developed countries [72]. With a few notable exceptions, such as the Arabian, Mongolian, and Icelandic horses, breeds (**Table 1**) have been created in the last two centuries [72]. The earliest horse studbook, that of the Thoroughbred racing horses, was created in 1791 [72]. The population structure resulting from selective breeding is characterized by high interbreed and low intrabreed genetic diversity [72]. Domestic horses exhibit remarkable variation in coat coloration, including the bay or bay-dun wild-type phenotypes, other basic colors like chestnut and black, as well as dilution (e.g., cream and silver), and spotting patterns (e.g., leopard complex, tobiano, and sabino) [72]. Horse locomotion has also been recurrently selected, including their ability to perform alternate gaits, such as four-beat, lateral, or diagonal ambling [72]. Although some horse breeds, such as the Thoroughbred racing horses, are still extremely popular, a significant part of this great diversity is currently endangered: 87 horse breeds are already extinct, and among the remaining 905, almost a quarter are categorized as at risk [72].

5. Final considerations

Ever since Darwin, the study of domestication has puzzled scientists [39]. Hundreds of articles are published each year [33, 39], as well as books, among which some are listed in **Table 2**. Despite this interest, both the words "domestication" and "domestic animal" remain confusing and poorly defined [32]. For domestication, this is mainly due to the inherent difficulty in assigning static terms to a process involving long-term and continuous change [32]. For "domestic animal," this is because this sort of dichotomous perspective wild/domestic is false and obscures the existence of transitional forms [32, 36, 73–75]. This is why the concept of "domestication level" was proposed for fish to describe more accurately the diversity of production methods as a continuum [52], from fishing up to the rearing of genetically improved animals [24, 25, 41, 76, 77]. This concept could be applied to other animals [26, 27] and may help describing the evolution of farmed species through both space and time in the future [36, 73].

Traditionally, the process of domestication was assumed to be initiated by humans, involving strong bottlenecks in the domestic population (corresponding to founder events due to the

Authors or editors	Date	Title	Ref.
Darwin CR	1859	On the Origin of Species by Means of Natural Selection, or the Preservation of Favoured Races in the Struggle for Life	[53]
Darwin CR	1868	The Variation of Animals and Plants under Domestication	[54]
Clutton-Brock J	1987	The Natural History of Domesticated Mammals	[55]
Digard JP	1990	L'homme et les animaux domestiques: Anthropologie d'une passion	[36]
Diamond J	1997	Guns, Germs, and Steel: The Fates of Human Societies	[3]
Guillaume J	2010	Ils ont domestiqué plantes et animaux: Prélude à la civilisation	[56]
Clutton-Brock J	2012	Animals as Domesticates: A World View Through History	[44]
Gepts et al.	2012	Biodiversity in Agriculture: Domestication, Evolution, and Sustainability	[57]
Vigne JD	2012	Les débuts de l'élevage	[58]
Wuerthner et al.	2014	Keeping the Wild: Against the Domestication of Earth	[59]
Francis RC	2015	Domesticated: Evolution in a Man-Made World	[60]
Alves and Albuquerque	2018	Ethnozoology: Animals In Our Lives	[61]
Scanes and Toukhsati	2018	Animals and Human Society	[62]

Table 2. Few examples of books focusing on animal domestications.

selection of only a few individuals at the beginning of the process) and reproductive isolation between wild and domestic forms [52, 67]. However, a growing body of archaeological, genetic, and ethnohistorical evidence suggests that long-term gene flow between wild and domestic stocks was much more common than previously expected, and selective breeding of females was largely absent during the early phases of animal domestication [52, 67]. Therefore, complete separation between wild and domestic populations was relatively late and region-specific [52]. These findings challenge assumptions about severe genetic bottlenecks during domestication and interpretations of genetic variability in terms of multiple instances of domestication and raise new questions regarding ways in which behavioral and phenotypic domestication traits were developed and maintained [52, 72]. The identity of the wild progenitor (or progenitors) of most domestic mammals remains also unclear because (i) the potential wild progenitors are often able to interbreed and produce fertile offspring with the domesticated congeners and (ii) many domestic animals can produce viable offspring with a host of wild, closely related sister taxa [32]. Therefore, the intuitive notion that each modern domestic animal (when discussed as a global population) is descended solely from a single wild species is almost certainly incorrect, and the genetic ancestry of domestics is likely to be relatively complex [32, 40].

Domesticated species are the result of a long and endless process that started millennia ago (**Table 1**). During about 98% of their domestication history, farm animals have been managed in a sustainable way by farmers, which lead to animals well adapted to local conditions [49, 50]. Yet, the situation changed dramatically 200 years ago as animals began to be selected for the same phenotypic characteristics to produce hundreds of well-defined breeds (**Table 1**), and reproduction among breeds was seriously reduced, leading to the fragmentation of the initial gene pool [49, 50, 70]. A few decades ago, the selection pressures were increased further, particularly with the use of artificial insemination, leading to a few industrial breeds with very high performances [49, 50, 70]. In the United States, the average milk production/cow of dairy cows increased by 1287 kg between 1993 and 2002, and 708 kg of this increase, or 55%, was due to genetics [78]. Interestingly, until the mid-1980s, most of the increase in milk yield was the result of improved management, in particular better application of nutritional standards and improved quality of rough age [78]. Since then, genetics became the major factor as a result of effective use of artificial insemination, intense selection based on progeny testing of bulls, and worldwide distribution of semen from bulls with high genetic merit for production [78]. This results in that, despite their total number of individuals, numerous industrial breeds have low effective population sizes [49, 50, 70]. This might explain that apart from a highly favorable increase in production, present-day selection for high production efficiency in livestock species in many cases was accompanied by undesirable side effects for several physiological, immunological, and reproduction traits [78, 79]. A new breeding goal aimed at improving fitness and tolerance of metabolic stress is necessary to prevent the decrease in the quality of life of farmed species and instead, perhaps, enhance it [70, 78–80]. More generally, an alternative to breeding for specific traits is to target "robustness" and "resilience," with the former focusing on current variation among environments and the latter on future variation [81]. Management strategies should be used to address short-term challenges from changing environments, and genetic selection should be used to address long-term problems [81]. Another solution might be to crossbreed domesticated animals either with their wild ancestor (if they still exist) or with wild relatives; it is therefore also urgent to properly assess the potential of the wild relatives as genetic resources for agriculture, and because most are endangered, actions should be implemented to preserve them [50]. Local breeds (present in only one country) in marginal areas are also seriously endangered [49, 50, 80]. For instance, in Europe more than 40% of livestock breeds are currently estimated to be endangered [82]. Farmers are often forced to abandon their traditional breeds and to raise more competitive industrial breeds [40, 83]. As a consequence, many locally adapted breeds have already disappeared [49, 50, 82]. Such a phenomenon can be very fast, and a valuable traditional breed can be lost within a decade [50]. Furthermore, even in less-developed countries, the introgression of genes from industrial breeds seriously compromises the long-term persistence of genetic resources in locally well-adapted breeds [49, 50, 83]. Adaptive traits may be rapidly lost by poorly designed crossbreeding, leading to dilution of important adaptive loci of traditional breeds. Traits such as resistance to local infectious and parasitic diseases, adaptation to poor forage, homing, and gregarious behavior can be rapidly lost and difficult to rescue [50]. According to the FAO, about 300 of 6000 breeds of farm animals have become extinct over the past 15 years, and 1350 currently face extinction in the near future [42, 50].

In conclusion, within a few decades, we might lose most of the highly valuable farm animal genetic resources that humans have gradually selected over the past millennia [45, 49, 50, 72]. Subsidies should therefore be urgently given to help farmers who contribute to the *in situ* preservation of genetic resources in marginal or rare breeds [80], and selection programs should aim at restoring the genetic diversity in industrial breeds [49, 50].

Conflict of interest

The author declares no conflict of interest.

Author details

Fabrice Teletchea

Address all correspondence to: fabrice.teletchea@univ-lorraine.fr

UR AFPA, Université de Lorraine, Vandoeuvre-Les-Nancy, France

References

[1] Larson G, Piperno DR, Allaby RG, Purugganan MD, Andersson L, Arroyo-Kalin M, et al. Current perspectives and the future of domestication studies. Proceedings of the National Academy of Sciences of the United States of America. 2014;**111**:6139-6146

[2] Mannion AM. Domestication and the origins of agriculture: An appraisal. Progress in Physical Geography. 1999;**23**:37-56

[3] Diamond J. Guns, Germs, and Steel: The Fates of Human Societies. New York: Norton.; 1997. 480 p

[4] Diamond J. Evolution, consequences and future of plant and animal domestication. Nature. 2002;**418**:700-707

[5] Vigne J-D. The origins of animal domestication and husbandry: A major change in the history of humanity and the biosphere. Comptes Rendus Biologies. 2011;**334**:171-181

[6] Vigne J-D. Early domestication and farming: What should we know or do for a better understanding? Anthropozoologica. 2015;**50**:123-150

[7] Boyd CE, AA MN. Aquaculture, Resource Use, and the Environment. Hoboken: John Wiley & Sons, Inc.; 2015. 337 p

[8] Roser M, Esteban Ortiz-Ospina E. World Population Growth. Published online at OurWorldInData.org. 2018. Retrieved from: https://ourworldindata.org/world-population-growth [Accessed: June 18, 2018]

[9] Foley JA, Ramankutty N, Brauman KA, Cassidy ES, Gerber JS, Johnston M, et al. Solutions for a cultivated planet. Nature. 2011;**478**:337342

[10] Duarte CM, Marba N, Holmer M. Rapid domestication of marine species. Science. 2007; **316**:382-383

[11] Price ARG. The marine food chain in relation to biodiversity. The Scientific World. 2001;**1**: 579-587

[12] Cury P, Cayré P. Hunting became a secondary activity 2 000 years ago: Marine fishing did the same in 2021. Fish and Fisheries. 2001;**2**:162-169

[13] Milner-Gulland EJ, Bennett EL, SCB 2002 Annual Meeting Wild Meat Group. Wild meat: The bigger picture. Trends in Ecology and Evolution. 2003;**18**:351-357

[14] Olden JD. Biotic homogenization: A new research agenda for conservation. Journal of Biogeography. 2006;**33**:2027-2039

[15] Teletchea F, Beisel JN. Alien fish species in France with emphasis on the recent invasion of gobies. In: Ray S, editor. Biological Resources of Water. 2018. pp. 75-92. DOI: 10.5772/intechopen.73408

[16] Shelton WL, Rothbard S. Exotic species in global aquaculture—A review. The Israeli Journal of Aquaculture—Bamidgeh. 2006;**58**:3-28

[17] De Silva SS, Nguyen TTT, Turchini GM, Amarasinghe US, Abery NW. Alien species in aquaculture and biodiversity: A paradox in food production. Ambio. 2009;**38**:24-28

[18] Subasinghe R, Soto D, Jia J. Global aquaculture and its role in sustainable development. Reviews in Aquaculture. 2009;**1**:2-9

[19] Pauly D. Anecdotes and the shifting baseline syndrome of fisheries. Trends in Ecology & Evolution. 1995;**10**:430

[20] Foley JA, DeFries R, Asner GP, Barford C, Bonan G, Carpenter SR, et al. Global consequences of land use. Science. 2005;**309**:570-574

[21] Ceballos G, Ehrlich PR, Dirzo R. Biological annihilation via the ongoing sixth mass extinction signaled by vertebrate population losses and declines. Proceedings of the National Academy of Sciences of the United States of America. 2017;**114**:E6089-E6096

[22] Cafaro P. Expanding parks, reducing human numbers, and preserving all the wild nature we can: A superior alternative to embracing the Anthropocene Era. In: Crist E, Butler T, editors. Wuerthner G, Keeping the Wild. Against the Domestication of Earth. Washington: Island Press; 2014. pp. 137-145

[23] Crutzen PJ. Geology of mankind. Nature. 2002;**415**:23

[24] Teletchea F. Qu'est-ce qu'un poisson domestique? Implications pour le développement futur de l'aquaculture. Ethnozootechnie. 2012;**90**:7-12

[25] Teletchea F, Fontaine P. Levels of domestication in fish: Implications for the sustainable future of aquaculture. Fish and Fisheries. 2014;**15**:181-195

[26] Teletchea F. Domestication and genetics: What a comparison between land and aquatic species can bring? In: Pontarotti P, editor. Evolutionary Biology: Biodiversification from Genotype to Phenotype, Chapter 20. Dordrecht: Springer; 2015. pp. 389-401

[27] Teletchea F. Wildlife Conservation: Is Domestication a Solution? In: Lameed GA, editor. Wildlife Conservation. Intech Editions; 2017. pp. 1-22. https://www.intechopen.com/books/global-exposition-of-wildlife-management

[28] Zeder MA. Pathways to animal domestication. In: Gepts P, Famula TR, Bettinger RL, et al., editors. Biodiversity in Agriculture: Domestication, Evolution, and Sustainability. Cambridge: Cambridge University Press; 2012

[29] Larson G, Burger J. A population genetics view of animal domestication. Trends in Genetics. 2013;**29**:197205

[30] Larson G, Fuller DQ. The evolution of animal domestication. Annual Review of Ecology, Evolution, and Systematics. 2014;**66**:115-136

[31] Zeder MA. Core questions in domestication research. Proceedings of the National Academy of Sciences of the United States of America. 2015;**112**:3191-3198

[32] Dobney K, Larson G. Genetics and animal domestication: New windows on an elusive process. Journal of Zoology. 2006;**269**:261-271

[33] Zeder MA. Central questions in the domestication of plants and animals. Evolutionary Anthropology. 2006;**15**:105-117

[34] Mignon-Grasteau S, Boissy A, Bouix J, Faure J-M, Fisher AD, Hinch GN, et al. Genetics of adaptation and domestication in livestock. Livestock Production Science. 2005;**93**:3-14

[35] Price EO. Behavioral development in animals undergoing domestication. Applied Animal Behaviour Science. 1999;**65**:245-271

[36] Digard J-P. L'homme et les animaux domestiques. Fayard: Anthropologie d'une passion; 2009. 325 p

[37] Trut L, Oskina I, Kharlamova A. Animal evolution during domestication: The domesticated fox as a model. BioEssays. 2009;**31**:349-360

[38] Price EO. Behavioral aspects of animal domestication. Quarterly Review Biology. 1984;**59**: 1-32

[39] Driscoll CA, MacDonald DW, O'Brien SJ. From wild animals to domestic pets, an evolutionary view of domestication. Proceedings of the National Academy of Sciences of the United States of America. 2009;**106**:9971-9978

[40] Groeneveld LF, Lenstra JA, Eding H, Toro MA, Scherf B, Pilling D, et al. Genetic diversity in farm animals—A review. Animal Genetics. 2010;**41**:6-31

[41] Teletchea F. Domestication of marine species: Update and perspectives. Journal of Marine Science and Engineering. 2015;**3**:1227-1243

[42] Scherf BD. World Watch List for Domestic Animal Diversity. 3rd ed. Vol. 2000. Rome: FAO; 2000. pp. 1-36

[43] Denis B. La domestication: Un concept devenu pluriel. INRA Productions Animales. 2004; **17**:161-166

[44] Cutton-Brock J. Animals as Domesticates. A World View through History. East Lansing: Michigan State University Press; 2012. 189 p

[45] Felius M, Beerling ML, Buchana DS, Theunissen B, Koolmees PA, Lenstra JA. On the history of cattle genetic resources. Diversity. 2014;**6**:705-750

[46] Wiener P, Wilkison S. Deciphering the genetic basis of animal domestication. Proceedings of the Royal Society B-Biological Sciences. 2011;**278**:3161-3170

[47] Verrier E, Saint-Dizier M. Contraintes et opportunités pour l'amélioration génétique des ruminants. INRA Productions Animales. 2011;**24**:297-306

[48] Woodford MH. World Watch List for Domestic Animal Diversity, Part 3. Wild Relatives of Domestic Livestock & Some Suggestions for New Domesticants. 3rd ed. Rome: FAO; 2000. pp. 647-716

[49] Taberlet P, Valentini A, Rezaei HR, Naderi S, Pompanon F, Negrini R, et al. Are cattle, sheep, and goats endangered species. Molecular Ecology. 2008;**17**:275-284

[50] Taberlet P, Coissac E, Pansu J, Pompanon F. Conservation genetics of cattle, sheep, and goats. Comptes Rendus Biologies. 2011;**334**:247-254

[51] Schilber J, Elsner J, Schlumbaum A. Incorporation of aurochs into a cattle herd in Neolithic Europe: Single event or breeding? Scientific Reports. 2014;**4**:5798

[52] Marshall FB, Dobney K, Denham T, Capriles JM. Evaluating the roles of directed breeding and gene flow in animal domestication. Proceedings of the National Academy of Sciences of the United States of America. 2014;**111**:6153-6158

[53] Darwin CR. On the Origin of Species by Means of Natural Selection, or the Preservation of Favoured Races in the Struggle for Life. London: John Murray; 1859. p. 502

[54] Darwin CR. The Variation of Animals and Plants Under Domestication. Vol. 1 and 2. London: John Murray; 1868. 411 and 486 p

[55] Clutton-Brock J. A Natural History of Domesticated Mammals. Cambridge: Cambridge University Press; 1987. 208 p

[56] Guillaume J. Ils ont domestiqué plantes et animaux. Prélude à la civilisation. Editions Quae; 2010. 453 p

[57] Gepts P, Famula TR, Bettinger RL, Brush SB, Damania AB, McGuire PE, et al. Biodiversity in Agriculture. Domestication, Evolution and Sustainability. Cambridge: Cambridge University Press; 2012. 606 p

[58] Vigne JD. Les débuts de l'élevage. Paris: Le Pommier/Universcience éditions; 2012. 189 p

[59] Wuerthner G, Crist E, Butler T, editors. Keeping the Wild. Against the Domestication of Earth. Washington: Island Press; 2014. 271 p

[60] Francis RC. Domesticated. Evolution in a Man-Made Evolution. New-York: WW Norton & Company; 2015. 484 p

[61] Alves RRN, Albuquerque UP, editors. Ethnozoology: Animals in Our Lives. London: Academic Press; 2018. 540 p

[62] Scanes CG, Toukhsati SR, editors. Animals and Human Society. London: Academic Press; 2018. 528 p

[63] De Roos APW, Hayes BJ, Spelman RJ, Goddard ME. Linkage disequilibrium and persistence of phase in Holstein-Friesian, Jersey and Angus cattle. Genetics. 2008;**179**:1503-1512

[64] Groenen MAM. A decade of pig genome sequencing: A window on pig domestication and evolution. Genetics Selection Evolution. 2016;**48**:23

[65] Chen K, Baxter T, Muir WM, Groenen MA, Schook LB. Genetic resources, genome mapping and evolutionary genomics of the pig (*Sus scrofa*). International Journal of Biological Sciences. 2007;**3**:153-165

[66] Wilkinson S, Lu ZH, Megens H-J, Archibald AL, Haley C, Jackson IJ, et al. Signatures of diversifying selection in European pig breeds. PLoS Genetics. 2013;**9**:e1003453

[67] Frantz LA, Schraiber J, Madsen O, Megens H-J, Cagan A, Bosse M, et al. Evidence of long-term gene flow and selection during domestication from analyses of Eurasian wild and domestic pig genome. Nature Genetics. 2015;**47**:1141-1149

[68] Bosse M. A genomics perspective on pig domestication. In: Teletchea F, editor. Animal Domestication. 2019. DOI: 10.5772/intechopen.82646. https://www.intechopen.com/online-first/a-genomics-perspective-on-pig-domestication

[69] Kijas JW, Lenstra JA, Hayes B, Boitard S, Porto Neto LR, et al. Genome-wide analysis of the world's sheep breeds reveals high levels of historic mixture and strong recent selection. PLoS Biology. 2012;**10**:e1001258

[70] Rodenburg TB, Turner SP. The role of breeding and genetics in the welfare of farm animals. Animal Frontiers. 2012;**2**:16-21

[71] Amills M, Capote J, Tosser-Klopp G. Goat domestication and breeding: A jigsaw of historical, biological and molecular data with missing pieces. Animal Genetics. 2017;**48**: 631-644

[72] Librado P, Fages A, Gaunitz C, Leonardi M, Wagner S, Khan N, et al. The evolutionary origin and genetic makeup of domestic horses. Genetics. 2016;**204**:423-434

[73] Digard JP. La domestication animale revisitée par l'anthropologie. Ethnozootechnie. 2003; **71**:33-44

[74] Lherminier P. Peut-on parler d'espèces domestiques. Ethnozootechnie. 2003;**71**:89-96

[75] Denis B. La conception zoologique classique de la domestication: Présentation générale. Ethnozootechnie. 2003;**71**:3-10

[76] Teletchea F. Fish domestication: An overview. In: Teletchea F, editor. Animal Domestication. 2019. DOI: 10.5772/intechopen.79628

[77] Teletchea F. Fish domestication in aquaculture: Reassessment and emerging questions. Cybium. 2019;**43**:7-15

[78] Oltenacu PA, Algers B. Selection for increased production and the welfare of dairy cows: Are new breeding goals needed? Ambio. 2005;**34**(4–5):311-315

[79] Raw WM, Kanis E, Noordhuizen-Stassen EN, Grommers FJ. Undesirable side effects of selection for high production efficiency in farm animals: A review. Livestock Production Science. 2008;**56**:15-33

[80] Boettcher PJ, Hoffmann I, Baumung R, Drucker AG, McManus C, Berg P, et al. Genetic resources and genomics for adaptation of livestock to climate change. Frontiers in Genetics. 2015;**5**:45-47

http://dx.doi.org/10.5772/intechopen.86783

[81] Cole JB, Bormann JM, Gill CA, Khatib H, Koltes JE, Maltecca C, et al. Breeding and genetics symposium: Resilience of livestock to changing environments. Journal of Animal Sciences. 2017;**95**:1777-1779

[82] Gandini GC, Villa E. Analysis of the cultural value of local livestock breeds: A methodology. Journal of Animal Breeding Genetics. 2003;**120**:1-1

[83] Biscarini F, Nicolazzi EL, Stella A, Boettcher PJ, Gandini G. Challenges and opportunities in genetic improvement of local livestock breeds. Frontiers in Genetics. 2015;**6**:178-184

Chapter 2

A Genomics Perspective on Pig Domestication

Mirte Bosse

Additional information is available at the end of the chapter

http://dx.doi.org/10.5772/intechopen.82646

Abstract

Land animal domestication has typically led to remarkable phenotypic diversity, stemming from a broad genetic background. The process of land animal domestication turns out to be a complex, long-term event with extensive gene-flow between wild and captive populations. Using pig as model, this chapter provides an in-depth overview of domestication-related events leading towards the genetic diversity in extant pig breeds. Five events in the evolutionary history and domestication of pigs can be recognized that are important for the genetic variation in modern pig genomes: (1) Speciation of Sus species in Island South-East Asia (ISEA); (2) Divergence between European and Asian lineages; (3) Independent domestication leading to separate domesticated clades in Europe and Asia; (4) Hybridization between domesticated pigs from Asia and Europe; and (5) Breed formation. Remarkably, the extensive mixture of genetic material leading towards the current European commercial pigs has resulted in domestic breeds that are genetically more diverse than their wild ancestors. Nowadays, commercial breeding and genomics go hand in hand. Genomics has not only proven useful to provide understanding about the domestication history of pigs but also about the molecular mechanisms underlying traits of interest. Moreover, genomic selection is an important tool integral to modern commercial breeding.

Keywords: genomics, pig, hybridization, selection, domestication

1. The process of livestock domestication

Domestication of land animals has typically led to a wide variety of domestic forms, with remarkable phenotypic diversity not seen in the wild. However, the underlying molecular variation resulting in a specific phenotype often stems from mutations predating domestication. Although domestication generally leads towards a reduction in effective population size, land animal domestication cannot be seen as a simple split of a subset of individuals from their wild progenitors. The meaning of the word domestication is poorly defined and lacks consistency across different scientific disciplines [1]. From a population genetics perspective, domestication results in a deliberate separation of the captive, and then domesticated population from its parent population. Domestication is, therefore, initially indistinguishable from any other event that results in reduction of gene-flow between populations, and creating opportunity to respond to new selective pressures [2]. The simplest definition of domestication considers a domestic population as a subset of the wild population with cessation of gene-flow [3]. Therefore, one can expect that domestication results in a reduction of genetic variation in the domesticated population. The onset of domestication occurred in multiple geographically distinct areas during the late Pleistocene to early Holocene transition (12,000–8200 B.P, [4]). The process of land animal domestication, however, turns out to be a complex, long-term event initiated by cultural transitions related to food production [5, 6]. The definition of an animal to be considered domesticated varies, however, some common characteristics emerge from literature. Teletchea and Fontaine propose that a domesticated animal should be selectively bred in captivity and modified from its wild ancestors [7]. It is important to realize that those early considered domestic populations were genetically and phenotypically hardly distinguishable from wild types, and therefore geographical location was a better predictor of local characteristics than domestication status [3]. The general assumption that multiple centers of domestication exist has important implications for the source of genetic and phenotypic variation in domesticated species. In cattle, for example, two distinct cattle lineages that separated ~300,000 ya, contributed to two major lineages of extant cattle, that is, taurine cattle (originating from *Bos taurus*) and indicine cattle (originating from *Bos indicus*) [8]. It is not unlikely that multiple populations of wild land animals that are now extinct contributed to the genetic diversity that is observed in modern breeds [9]. The domestic animal populations accompanying human settlements did not necessarily remain at their original location of domestication. Rather, they moved along with early farmers spreading in Asia and from Eastern Anatolia throughout Europe [10]. During this process, the connection of domestic animals and farmers was relatively loose, enabling animals to hybridize with local wild populations [11]. Only centuries later, animals were actually kept in strict enclosures and intentionally bred for specific purposes, leading towards the best-known characteristic of domestic animals: docility [12]. This controlled environment drastically reduced the opportunity of domestic herds to interbreed with local wild populations, which enabled strong divergence between domestic and wild forms. We should realize the genetic basis of the modifications leading towards morphological differences in domestic animals compared to their wild ancestors is mostly provided by standing genetic variation, that is, mutations that were already present before the onset of domestication and selection. Therefore, indicating the genetic underpinnings of domestication remain challenging [13, 14]. Arguably, we can speak about a domestic population if not only the gene pool is distinct from the wild variety, but also (artificially) selected variants leading to desired phenotypes are at high(er) frequency in the domestic population [15–18]. In this chapter, an in-depth overview is provided for the complex process of domestication, admixture, and selection leading towards the genetic diversity in extant breeds, using pig as model.

2. Genomic insight in pig domestication

Domesticated species are good models to study genomic and phenotypic consequences of demography and selection [19]. The use of higher DNA marker densities has enabled researchers to reveal the complexity of livestock domestication, which was shown to be far more complex than a single sampling from the wild [20]. Genotyping and sequencing technologies have opened up many opportunities to reveal the complex history of domestication, admixture, and selection in livestock [4, 20]. Combining modern sequence technologies with extensive studies on fossil records and land animal usage now enables the reconstruction of domestication in details. Apart from a suitable history and documentation, the availability of detailed genetic information is crucial to be able to study genomic alterations due to domestication. Pig (*Sus scrofa*, Linnaeus, 1758) was the first livestock species for which a genome consortium was established with the intention to completely map the genome [21, 22]. The design of a 60k single nucleotide polymorphism (SNP) chip for pigs in 2009 greatly contributed to the applicability of genomics techniques in pig breeding, and simultaneously increased possibilities for population genomics studies [23]. The establishment of a consortium to sequence the pig genome in 2003 and publication of the pig reference genome in 2012 opened up an even greater window of opportunities to study various aspects of the genetics of pig, since the highest resolution possible became reality [21, 24]. Together with the evolutionary history of pig, these provide an unprecedented study system to demonstrate the impact of domestication from a genomics perspective. Pig genomes contain a complex composition of segments, reflecting the different backgrounds that contributed to the domestic animal it is today. Disentangling these genomic signatures provides enormous information about the complex background and history of the worlds' most consumed meat type [25].

3. Conceptual history of the pig (*Sus scrofa*)

Here I will discuss genomic variation within and between different populations of pigs, providing deeper understanding of how domestication has influenced genetic diversity of pigs. Five major events in the evolutionary history and domestication of pigs can be recognized that are of importance for the distribution of genetic variation in modern pig genomes (**Figure 1**).

3.1. Speciation of *Sus* in island South-East Asia

Knowledge about the source of the domesticated form, the origin of the species, is essential to understand genetic variation within modern breeds. The Suidae family is particularly interesting for molecular genetic studies as it is one of the few mammalian lineages that has closely related species living today. Multiple *Sus* species originated roughly ~4 million years ago on Island Southeast Asia (ISEA). The island structure in this region probably promoted speciation, since the bearded pig *Sus barbatus*, the warty pigs *S. celebensis* and *S. verrucosus* but also wild *S. scrofa* occur on separate islands. The phylogenetic structure within the genus *Sus*

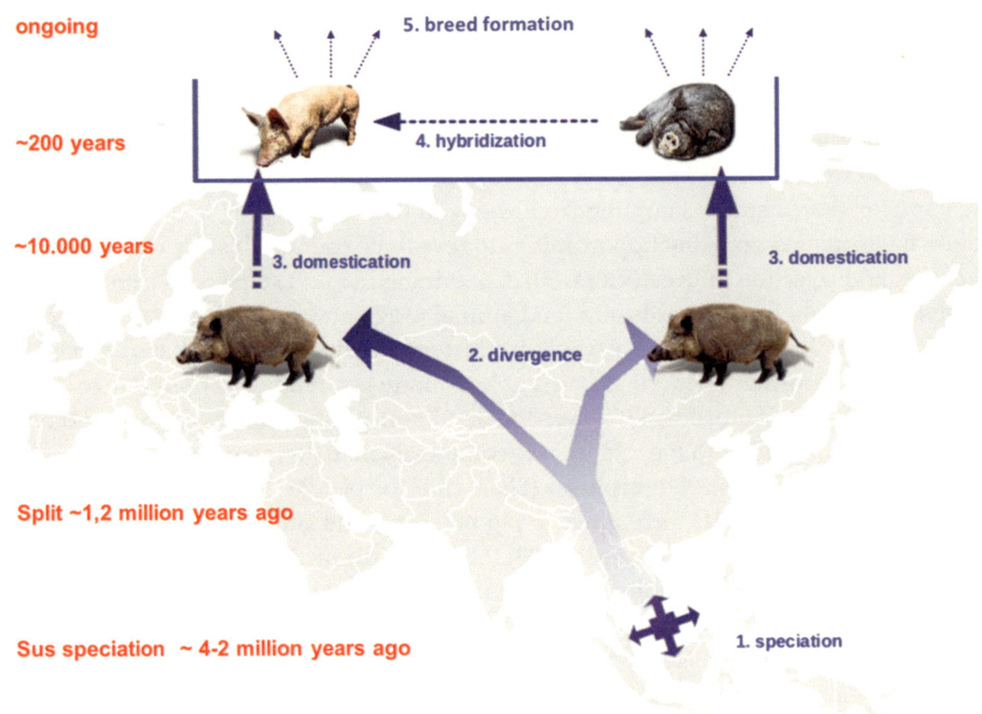

Figure 1. Schematic overview of the history of the pig (*Sus scrofa*). Five main events in pig history are indicated in blue, with the approximate timing of those events in red: (1) Speciation of *Sus* species in Island South-East Asia (ISEA). (2) Divergence between European and Asian *S. scrofa* lineages. (3) Independent domestication leading to separate domesticated clades in Europe and Asia. (4) Hybridization between domesticated pigs from Asia and Europe. (5) Breed formation.

has been studied intensively and revealed a complex history of admixture [26, 27]. The past connection of landmasses at the Sunda shelf and isolation of Indonesian islands by the rapid sea level rise after the last glaciation period [28] created a dynamic process of (re)colonization, isolation and admixture of different *Sus* species and populations [29, 30]. The species that gave rise to the domesticated pig, *Sus scrofa*, has its origin in Southeast Asia some ~4 Mya and colonized almost the entire Eurasian mainland from there. The widespread and opportunistic nature of this species probably contributed to the fact that *Sus scrofa* is the only pig species that was successfully domesticated [25].

3.2. Divergence between European and Asian *S. scrofa*

Sus scrofa is widespread within Eurasia (**Figure 1**) and consists of many isolated wild and domesticated populations. The divergence between Western-European and Eastern-Asian populations has been estimated at about 1.2 Mya [24, 29], and has resulted in many fixed molecular differences between the two groups. This divergence not only resulted in a European and an Asian *S. scrofa* clade, but also in differences in demographic history and population size. The last glacial maximum probably reduced population sizes of both European and Asian wild boars, but the reduction was most severe in Europe [24]. The geographic distribution of wild boar over Europe faced another severe decline starting in the middle ages and lasted until the late eighteenth century [25]. In the mid-nineteenth century, natural or human-mediated recolonization events resulted in isolated populations expanding their range. Some of these isolated populations were small in effective size for decades or longer, causing inbreeding and population differentiation. Local re-stocking of populations with geographically distinct wild boar resulted in complex genetic structures and signatures of population dynamics [31, 32]. Such complex genetic architectures have been detected in Italian and Luxembourgian wild boars. However, these mixed genomes could have been shaped due to ancient glaciation events [33] or because of recent mixture [34]. Asian *Sus scrofa* is thought to have had a larger effective population size which, together with its proximity to the origin of the species, results in higher genetic diversity compared to the European clade [24, 35, 36]. These highly distinct groups of wild boar provided the basis of the genetic background of the later domesticated pigs.

3.3. Independent domestication leading to separate clades

The demographic and geographic history of the domesticated pig may be just as complex as that of its wild counterpart. There is compelling evidence that pig domestication events occurred at multiple locations, Eastern Anatolia and China independently, some 9000–10,000 years ago [26, 37]. Domestication has not been a single event, but rather a long period with recurrent admixture with wild populations [38]. Following initial domestication, the traits selected as well as how animals were kept, strongly differed in Europe and Asia resulting in highly different domesticated pigs between Europe and Asia. Asian pigs were kept in close proximity of humans, often integrated in their settlements. By contrast, European pigs were roaming freely in forested areas in the surroundings [39, 40]. Only during the Industrial Revolution, a more strict pig farming system was adopted and implemented to fulfill the increasing demand for pork. Because of recurrent gene-flow between wild and domestic pigs, a reduction in genetic diversity cannot be observed in domesticated pigs compared to their presumed wild counterparts [35, 38, 41]. One should realize though that European and Asian domesticated pigs have been geographically isolated for over a million years ago, because they have distinct wild origins. Therefore, they genetically resemble local wild boar more than domestic pigs from different geographic origins [24, 35]. This dichotomy also underlies the fact that European pigs and wild boar are genetically less diverse than Asian wild boar and domestic pigs.

3.4. Hybridization between domesticated pigs of different origin

It is well documented that during the Industrial Revolution in Europe, European pigs have been deliberately hybridized with Asian pigs. Urbanization in Europe increased the demand for meat such as pork, but during those times, pig farmers would still have their pigs roaming in surrounding forests. Forest cover was decreasing and a different pig production system seemed inevitable [40]. Due to this changing environment, pig breeders sought a way to improve their stock in such way that pigs had to become adapted to living in small(er) enclosures, be more prolific and gain weight more rapidly. This led to selection for traits better adapted to the changed environment. Many of these traits were already present in Asian domestic pigs. Therefore, British farmers started crossbreeding their own pigs with these Asian pigs [40]. This introgression of Asian genetic material into European populations has long been demonstrated by genetic markers [42, 43]. Moreover, the intentional crossbreeding and consecutive artificial selection on Asia-derived traits enabled adaptive loci to emerge in the genome of European domestic pigs. Genes of Asian origin have been demonstrated to contribute to increased fertility and fatness in commercial Large White pigs [44, 45]. Very recently, hybridization between wild and domesticated pigs has been reported in Western Europe, resulting in traceable Asian genetic material in local wild boar populations in Germany [31, 32, 34].

3.5. Breed formation and globalization

Due to the worldwide consumption of pork, the species is farmed at a global scale, far exceeding its original natural distribution (IUCN). The influence and contribution of commercial pig breeds to local ecology and biodiversity is however debated [31, 32, 46]. Also, escape or intentional release of local stocks have resulted in feralization of domesticated pigs, which is now a major population in the United States, although the continent is not part of the native range of the species [47]. The domesticated pig as it is used nowadays for agricultural purposes consists of many breeds that have been separated and kept isolated for decades, which has resulted in many genetic differences between these breeds. Breed and population specific genetic studies have greatly enhanced the dissection of complex traits that are economically important. Knowing and understanding the origin and distribution of variation in (domesticated) species is important for conservation of genetic resources, such as culturally important heritage breeds [48]. Local husbandry and breeding techniques have created an enormous diversification of pig breeds. Generally, European breeds can be categorized into global commercial breeds, stemming from the White type in England, and local heritage breeds, developed locally and now often endangered [39]. It is notable that many heritage breeds genetically resemble the local wild boar more than global pig breeds, most likely because they were not improved by Asian gene-flow two centuries ago [35, 49–51]. The globalization of pig breeding and consumption has swamped local pig breeds with common commercial breeds from British heritage background, such as Large White, Landrace, Pietrain and Duroc [39]. Also, extensive admixture between breeds of different origin is known to occur, highly dependent on local breeding practices.

4. The hybrid nature of (pig) genomes

Increasing evidence showed that humans play an important role in stimulating hybridization in wild species, either unintentionally or on purpose. Human-induced hybridization can not only be a by-product of globalization as some species became widely distributed due to human mobility, but it can also be intentional such as in domesticated species [40, 52]. It is becoming apparent that many livestock species/breeds are actually a mixture of highly divergent populations with a mixed demographic history, combined in one genome. The formation of livestock breeds provides a good example of how man has influenced the genomic architecture of a species. In cattle, for example, exchange of genetic material between different species promoted the uptake of beneficial traits from closely related species [53]. In pig, domestication does not seem to have left a clear population bottleneck, as demonstrated by the high level of genetic variation in European pigs [38]. This suggests that the majority of the genetic variation that is present in European wild boar is also present in domestic breeds, even though modern pigs are phenotypically clearly different from their wild counterparts. Moreover, the gene-flow with wild populations as well as between different domestic lineages enabled pig breeders to select for locally and globally preferred traits, using a broad genetic background [44, 45]. Remarkably, the extensive mixture of genetic material leading towards the current European commercial pigs has resulted in domestic breeds that are genetically more diverse than their wild ancestors in Europe [24, 35, 36, 41]. This counter-intuitive characteristic of commercial pigs is mainly driven by the influx of Asian genes during the Industrial Revolution [45]; local heritage breeds that do not display signs of Asian gene-flow tend to have lower genetic diversity [50, 51]. Nowadays, many breeds and definitions are used to describe the origin of (local) stock, with some being a complex mixture of Asian and European heritage, depending on the geographical region and the breeding practice of pig farmers.

5. Breeding and genomics go hand in hand

Pig farming has drastically changed since first domestication. Todays' elaborate pig breeding industry has only few characteristics in common with early pig farmers, and has resulted in a highly professional large-scale pork production system, making use of latest technologies in animal breeding. Selection for particular traits not only improved due to more precise phenotyping and better defined traits such as carcass quality, growth rate and fertility [54], but also because of crossing breeds with desirable traits of different origins [45]. The use of pedigree information and large-scale tracking of animal relatedness has speeded up the improvement of pig breeds. In other livestock, especially cattle, the implementation of the use of genetic markers on top of pedigree information resulted in even more efficient selection [55]. The recent and rapid genetic progress can be achieved due to the implementation of genomic selection, in which animals are selected based on their performance predicted from their genotypes, rather than phenotypes [56]. This way, animals can be selected at an earlier stage, and predicted phenotypes for typically female traits can also be implemented using genotype information from males [57, 58].

Genomics has not only proven useful as a tool in genomic selection, but also has provided more understanding about the molecular mechanisms that underlie traits of interest. Knowledge about the link between genes and trait enables more accurate breeding [54]. Moreover, if the function of a specific gene is known, it can provide insight into the selection history of a breed. Numerous studies have successfully identified selection for genes linked to specific commercially important traits (**Table 1**). Interestingly, some of these genes under selection in European breeds have an Asian origin [59–61]. Also, genome-wide scans for detrimental variants have identified mutations in commercial populations with negative effects [62, 63].

Gene	Trait	Study
KIT	Coat color	Andersson and Plastow [66]
KITLG	Coat color	Okumura et al., [61]
MC1R	Coat color	Kijas et al., [67]; Fang et al., [68]
EDNRB	Coat color	Ai et al., [59]; Wilkinson et al., [69]
IGF2	Lean growth	van Laere et al., [70]
RYR1	Lean growth	Fujii et al., [71]
PRKAG3	Lean growth	Milan et al., [72]
NR6A1	Body size	Rubin et al., [73]
PLAG1	Body size	Rubin et al., [73]
LCORL	Body size	Rubin et al., [73]
OSTN	Body composition	Rubin et al., [73]
CLDN1	Fertility	Choi et al., [74]
AHR	Fertility	Bosse et al., [44]
TWIST1	Fatness	Choi et al., [74]
LEMD3	Ear morphology	Wilkinson et al., [59]

Table 1. Non-exhaustive list of genes associated with commercially important traits in pigs.

Recent work demonstrates that some variants that cause lethality in homozygous state are present at relatively high frequency in commercial pig lines [64, 65]. Knowing these recessive lethal mutations can aid in avoiding matings between carriers of such mutations within the breeding scheme. Overall, genomics has provided valuable insight into variation in pigs: what its origin is, how is it maintained, reduced and increased. This turned out to be a complex interplay of molecular processes, selection, demographic history, gene-flow and human interference. Moreover, genomics is an important tool in the pig industry nowadays and is integral to modern commercial breeding.

Author details

Mirte Bosse[1,2*]

*Address all correspondence to: mirte.bosse@wur.nl

1 Wageningen University & Research – Animal Breeding and Genomics, Wageningen, The Netherlands

2 VU University Amsterdam – Animal Ecology, Amsterdam, The Netherlands

References

[1] Zeder MA. The domestication of animals. Journal of Anthropological Research. 2012; **68**:161-190

[2] Zeder MA. Domestication as a model system for the extended evolutionary synthesis. Interface Focus. 2017;**7**:20160133. DOI: 10.1098/rsfs.2016.0133

[3] Larson G, Burger J. A population genetics view of animal domestication. Trends in Genetics. 2013;**29**(4):197-205

[4] Larson G, Piperno DR, Allaby RG, Purugganan MD, Andersson L, Arroyo-Kalin M, et al. Current perspectives and the future of domestication studies. Proceedings of the National Academy of Sciences of the United States of America. 2014;**111**(17):6139-6146

[5] Gepts P et al., editors. Biodiversity in Agriculture: Domestication, Evolution, and Sustainability. Cambridge: Cambridge Univ Press; 2012

[6] Price D, Bar-Yosef O, editors. The Beginnings of Agriculture: New Data, New Ideas. Curr Anthropol 52 (Supp 4). Chicago: Wenner-Gren Symposium Series; Univ of Chicago Press; 2011

[7] Teletchea F, Fontaine P. Levels of domestication in fish: Implications for the sustainable future of aquaculture. Fish and Fisheries. 2014;**15**:181-195

[8] Pitt D, Sevane N, Nicolasi E, MacHugh D, Colli L, Martinez R, et al. Domestication of cattle: Two or three events? Evolutionary Applications. 2018;1-14. DOI: 10.1111/eva.12674

[9] Scheu A, Powell A, Bollongino R, Vigne J-D, Tresset A, Çakırlar C, et al. The genetic prehistory of domesticated cattle from their origin to the spread across Europe. BMC Genetics. 2015;**16**:54. DOI: 10.1186/s12863-015-0203-2

[10] Driscoll CA, Macdonald DW, O'Brien SJ. From wild animals to domestic pets, an evolutionary view of domestication. Proceedings of the National Academy of Sciences of the United States of America. 2009;**106**(suppl. 1):9971. DOI: 10.1073/pnas.0901586106

[11] Zeder MA, Emshwiller E, Smith BD, Bradley DG. Documenting domestication: The intersection of genetics and archaeology. Trends in Genetics. 2006;**22**:139-155

[12] Zeder MA. Central questions in the domestication of plants and animals. Proceedings of the National Academy of Sciences of the United States of America. 2006;**15**:105-117

[13] Christie MR, Marine ML, French RA, Blouin MS. Genetic adaptation to captivity can occur in a single generation. Proceedings of the National Academy of Sciences of the United States of America. 2012;**109**:238-242

[14] Christie MR, Marine ML, Fox SE, French RA, Blouin MS. A single generation of domestication heritably alters the expression of hundreds of genes. Nature Communications. 2016;**7**:10676. DOI: 10.1038/ncomms10676

[15] Diamond J. Evolution, consequences and future of plant and animal domestication. Nature. 2002;**418**:700-707

[16] Mignon-Grasteau S, Boissy A, Bouix J, et al. Genetics of adaptation and domestication in livestock. Livestock Science. 2005;**93**:3-14

[17] Mirkena T, Duguma G, Haile A, Tibbo M, Okeyo AM, Wurzinger M, et al. Genetics of adaptation in domestic farm animals: A review. Livestock Science. 2010;**132**:1-12

[18] Teletchea F. Domestication and genetics: What a comparison between land and aquatic species can bring. In: Pontarotti P, editor. Evolutionary Biology. Biodiversification from Genotype to Phenotype. Cham, Switzerland: Springer; 2015. pp. 389-402

[19] Megens HJ, Groenen MAM. Domestication: A long-term genetic experiment domesticated species form a treasure-trove for molecular characterization of Mendelian traits by exploiting the specific genetic structure of these species in across-breed genome wide association studies. Heredity. 2012;**109**(1):1-3

[20] Bruford MW, Bradley DG, Luikart G. DNA markers reveal the complexity of livestock domestication. Nature Reviews. Genetics. 2003;**4**(11):900-910

[21] Schook LB, Beever JE, Rogers J, Humphray S, Archibald A, Chardon P, et al. Swine genome sequencing consortium (SGSC): A strategic roadmap for sequencing the pig genome. Comparative and Functional Genomics. 2005;**6**:251-255

[22] Archibald AL et al. Pig genome sequence – Analysis and publication strategy. BMC Genomics. 2010;**11**:438

[23] Ramos AM, Crooijmans RP, Affara NA, Amaral AJ, Archibald AL, Beever JE, et al. Design of a high density SNP genotyping assay in the pig using SNPs identified and characterized by next generation sequencing technology. PLoS One. 2009;**4**(8):e6524

[24] Groenen MA, Archibald AL, Uenishi H, Tuggle CK, Takeuchi Y Rothschild MF Rogel-Gaillard C, Park C Milan D Megens HJ, et al. Analyses of pig genomes provide insight into porcine demography and evolution. Nature. 2012;**491**(7424):393-398

[25] Groenen MAM. A decade of pig genome sequencing: A window on pig domestication and evolution. Genetics Selection Evolution. 2016;**48**:23. DOI: 10.1186/s12711-016-0204-2

[26] Larson G, Dobney K, Albarella U, Fang M, Matisoo-Smith E, Robins J, et al. Worldwide phylogeography of wild boar reveals multiple centers of pig domestication. Science. 2005;**307**:1618-1621

[27] Mona S, Randi E, Tommaseo-Ponzetta M. Evolutionary history of the genus Sus inferred from cytochrome b sequences. Molecular Phylogenetics and Evolution. 2007;**45**(2):757-762

[28] Hanebuth T, Stattegger K, Grootes PM. Rapid flooding of the Sunda shelf: A late-glacial sea-level record. Science. 2000;**288**(5468):1033-1035

[29] Frantz LA, Schraiber JG, Madsen O, Megens HJ, Bosse M, Paudel Y, et al. Genome sequencing reveals fine scale diversification and reticulation history during speciation in Sus. Genome Biology. 2013;**14**(9):R107

[30] Frantz LAF, Madsen O, Megens H-J, Groenen MAM, Lohse K. Testing models of speciation from genome sequences: Divergence and asymmetric admixture in island south-east Asian Sus species during the Plio-Pleistocene climatic fluctuations. Molecular Ecology. 2014;**23**:5566-5574

[31] Goedbloed DJ, Megens HJ, van Hooft P, Herrero-Medrano JM, Lutz W, Alexandri P, et al. Genome-wide single nucleotide polymorphism analysis reveals recent genetic introgression from domestic pigs into northwest European wild boar populations. Molecular Ecology. 2013;**22**(3):856-866

[32] Goedbloed DJ, van Hooft P, Megens HJ, Langenbeck K, Lutz W, Crooijmans RPMA, et al. Reintroductions and genetic introgression from domestic pigs have shaped the genetic population structure of northwest European wild boar. BMC Genetics. 2013;**14**:43

[33] Scandura M, Iacolina L, Crestanello B, Pecchioli E, Di Benedetto MF, Russo V, et al. Ancient vs. recent processes as factors shaping the genetic variation of the European wild boar: Are the effects of the last glaciation still detectable? Molecular Ecology. 2008;**17**(7):1745-1762

[34] Frantz AC, Zachos FE, Kirschning J, Cellina S, Bertouille S, Mamuris Z, et al. Genetic evidence for introgression between domestic pigs and wild boars (*Sus scrofa*) in Belgium and Luxembourg: A comparative approach with multiple marker systems. Biological Journal of the Linnean Society. 2013;**110**(1):104-115

[35] Bosse M, Megens HJ, Madsen O, Frantz LAF, Paudel Y, Crooijmans RPMA, et al. Untangling the hybrid nature of modern pig genomes: A mosaic derived from biogeographically distinct and highly divergent *Sus scrofa* populations. Molecular Ecology. 2014;**23**(16):4089-4102

[36] Bosse M, Madsen O, Megens HJ, Frantz LA, Paudel Y, Crooijmans RP, et al. Hybrid origin of European commercial pigs examined by an in-depth haplotype analysis on chromosome 1. Frontiers in Genetics. 2014;**5**:442

[37] Megens HJ, Crooijmans RPMA, Cristobal MS, Hui X, Li N, Groenen MAM. Biodiversity of pig breeds from China and Europe estimated from pooled DNA samples: Differences in microsatellite variation between two areas of domestication. Genetics Selection Evolution. 2008;**40**(1):103-128

[38] Frantz LA, Schraiber JG, Madsen O, Megens HJ, Cagan A, Bosse M, et al. Evidence of long-term gene flow and selection during domestication from analyses of Eurasian wild and domestic pig genomes. Nature Genetics. 2015;**47**:1141-1148

[39] Porter V. Pigs – A Handbook to the Breeds of the World. Mountfield, East Sussex: Helm Information; 1993

[40] White S. From globalized pig breeds to capitalist pigs: A study in animal cultures and evolutionary history. Environmental History. 2011;**16**(1):94-120

[41] Bosse M, Megens HJ, Madsen O, Paudel Y, Frantz LA, Schook LB, et al. Regions of homozygosity in the porcine genome: Consequence of demography and the recombination landscape. PLoS Genetics. 2012;**8**(11):e1003100

[42] Giuffra E, Kijas JMH, Amarger V, Carlborg O, Jeon JT, Andersson L. The origin of the domestic pig: Independent domestication and subsequent introgression. Genetics. 2000; **154**(4):1785-1791

[43] Amaral AJ, Ferretti L, Megens HJ, Crooijmans RPMA, Nie HS, Ramos-Onsins SE, et al. Genome-wide footprints of pig domestication and selection revealed through massive parallel sequencing of pooled DNA. PLoS One. 2011;**6**(4):e14782

[44] Bosse M, Megens HJ, Frantz LAF, Madsen O, Larson G, Paudel Y, et al. Genomic analysis reveals selection for Asian genes in European pigs following human-mediated introgression. Nature Communications. 2014;**5**:4392

[45] Bosse M, Lopes MS, Madsen O, Megens HJ, Crooijmans RP, Frantz LA, et al. Artificial selection on introduced Asian haplotypes shaped the genetic architecture in European commercial pigs. Proceedings of the Biological Sciences. 2015;**282**:pii:20152019. DOI: 10.1098/rspb.2015.2019

[46] Hall SJG, Bradley DG. Conserving livestock breed biodiversity. Trends in Ecology & Evolution. 1995;**10**(7):267-270

[47] Mayer JJ, Brisbin IL Jr. Wild Pigs in the United States: Their History, Comparative Morphology, and Current Status. 2nd ed. Athens, GA: The University of Georgia Press; 2008

[48] Groeneveld LF, Lenstra JA, Eding H, Toro MA, Scherf B, Pilling D, et al. Genetic diversity in farm animals - a review. Animal Genetics. 2010;**41**:6-31

[49] Herrero-Medrano JM, Megens HJ, Crooijmans RP, Abellaneda JM, Ramis G. Farm-by-farm analysis of microsatellite, mtDNA and SNP genotype data reveals inbreeding and crossbreeding as threats to the survival of a native Spanish pig breed. Animal Genetics. 2013;**44**(3):259-266

[50] Herrero-Medrano JM, Megens HJ, Groenen MAM, Bosse M, Perez-Enciso M, Crooijmans RPMA. Whole-genome sequence analysis reveals differences in population management and selection of European low-input pig breeds. BMC Genomics. 2014;**15**:601

[51] Herrero-Medrano JM, Megens HJ, Groenen MAM, Ramis G, Bosse M, Perez-Enciso M, et al. Conservation genomic analysis of domestic and wild pig populations from the Iberian Peninsula. BMC Genetics. 2013;**14**:106

[52] Kijas JW, Lenstra JA, Hayes B, Boitard S, Porto Neto LR, et al. Genome-wide analysis of the world's sheep breeds reveals high levels of historic mixture and strong recent selection. PLoS Biology. 2012;**10**:e1001258

[53] Wu D-D, Ding X-D, Wang S, Wojcik JM, Zhang Y, Tokarska M, et al. Pervasive introgression facilitated domestication and adaptation in the Bos species complex. Nature Ecology & Evolution. 2018;**2**:1139-1145

[54] Merks J, Mathur P, Knol E. New phenotypes for new breeding goals in pigs. Animal. 2012;**6**(04):535-543

[55] Doekes HP, Veerkamp RF, Bijma P, Hiemstra SJ, Windig JJ. Trends in genome-wide and region-specific genetic diversity in the Dutch-Flemish Holstein–friesian breeding program from 1986 to 2015. Genetics, Selection, Evolution. 2018;**50**:15. DOI: 10.1186/s12711-018-0385-y

[56] Meuwissen T, Hayes B, Goddard M. Prediction of total genetic value using genome-wide dense marker maps. Genetics. 2001;**157**(4):1819-1829

[57] Hayes BJ, Bowman PJ, Chamberlain AC, Goddard ME. Genomic selection in dairy cattle: Progress and challenges. Journal of Dairy Science. 2008;**92**:433-443

[58] Hayes BJ, Visscher PM, Goddard ME. Increased accuracy of selection by using the realised relationship matrix. Genetical Research. 2009;**91**:47-60

[59] Wilkinson S, Lu ZH, Megens HJ, Archibald AL, Haley C, Jackson IJ, et al. Signatures of diversifying selection in European pig breeds. PLoS Genetics. 2013;**9**(4):e1003453

[60] Ojeda A, Huang LS, Ren J, Angiolillo A, Cho IC, Soto H, et al. Selection in the making: A worldwide survey of haplotypic diversity around a causative mutation in porcine IGF2. Genetics. 2008;**178**(3):1639-1652

[61] Okumura N, Matsumoto T, Hamasima N, Awata T. Single nucleotide polymorphisms of the KIT and KITLG genes in pigs. Animal Science Journal. 2008;**79**(3):303-313

[62] Bosse M, Megens HJ, Derks MFL, de Cara MAR, Groenen MAM. Deleterious alleles in the context of domestication, inbreeding and selection. Evolutionary Applications. 2018:1-12. doi.org/10.1111/eva.12691

[63] Makino T, Rubin CJ, Carneiro M, Axelsson E, Andersson L, Webster MT. Elevated proportions of deleterious genetic variation in domestic animals and plants. Genome Biology and Evolution. 2018;**10**(1):276-290

[64] Derks MFL, Megens HJ, Bosse M, Lopes MS, Harlizius B, Groenen MAM. A systematic survey to identify lethal recessive variation in highly managed pig populations. BMC Genomics. 2017;**18**:858

[65] Derks MFL, Lopes MS, Bosse M, Madsen O, Dibbits B, Harlizius B, et al. Balancing selection on a recessive lethal deletion with pleiotropic effects on two neighboring genes in the porcine genome. PLoS Genetics. 2018;**14**(9):e1007661. DOI: 10.1371/journal.pgen.1007661

[66] Andersson L, Plastow G. Molecular genetics of coat colour variation. In: Rothschild MF, Ruvinsky A, editors. The Genetics of the Pig. 2nd ed. Wallingford: CABI; 2011. pp. 38-50

[67] Kijas JMH, Wales R, Tornsten A, Chardon P, Moller M, Andersson L. Melanocortin receptor 1 (MC1R) mutations and coat color in pigs. Genetics. 1998;**150**(3):1177-1185

[68] Fang M, Larson G, Ribeiro HS, Li N, Andersson L. Contrasting mode of evolution at a coat color locus in wild and domestic pigs. PLoS Genetics. 2009;**5**(1):e1000341

[69] Ai H, Fang X, Yang B, Huang Z, Chen H, Mao L, et al. Adaptation and possible ancient interspecies introgression in pigs identified by whole-genome sequencing. Nature Genetics. 2015;**47**:217-225

[70] Van Laere AS, Nguyen M, Braunschweig M, Nezer C, Collette C, Moreau L, et al. A regulatory mutation in IGF2 causes a major QTL effect on muscle growth in the pig. Nature. 2003;**425**:832-836

[71] Fujii J, Otsu K, Zorzato F, de Leon S, Khanna VK, Weiler JE, et al. Identification of a mutation in porcine ryanodine receptor associated with malignant hyperthermia. Science. 1991;**253**:448-451

[72] Milan D, Jeon JT, Looft C, Amarger V, Robic A, Thelander M, et al. A mutation in PRKAG3 associated with excess glycogen content in pig skeletal muscle. Science. 2000; **288**:1248-1251

[73] Rubin CJ, Megens HJ, Barrio AM, Maqbool K, Sayyab S, Schwochow D, et al. Strong signatures of selection in the domestic pig genome. Proceedings of the National Academy of Sciences of the United States of America. 2012;**109**(48):19529-19536

[74] Choi JW, Chung WH, Lee KT, Cho ES, Lee SW, Choi BH, et al. Whole-genome resequencing analyses of five pig breeds, including Korean wild and native, and three European origin breeds. DNA Research. 2015;**22**:259-267

Chapter 3

Insects: The Disregarded Domestication Histories

Thomas Lecocq

Additional information is available at the end of the chapter

http://dx.doi.org/10.5772/intechopen.81834

Abstract

Domestication has irrevocably impacted human evolution. The domestication process/pathways have been the focus of abundant research for plants and vertebrates. Advances in genetics and archeology have allowed tremendous progresses in the understanding of domestication for these organisms. In contrast, insects' domestication has comparatively received far less attention to date. Yet, insects are the most common animal group on Earth and provide many valuable ecosystem services to humans. Therefore, the aims of this chapter are (i) to provide an overview of main ancient and recent insect domestication histories and (ii) to reread them by the light of the domestication process, pathways, triggers, and consequences observed in other animal species. Some of the considered species (i.e., silkworm and honey bee) have been chosen because they are among the few insects commonly acknowledged as domesticated, while others allow illustrating alternative domestication patterns. The overview of current literature shows similar human-directed pathway and domestication syndrome (e.g., increased tameness, decreased aggressiveness, modified reproduction) between several insect species.

Keywords: domestication level, domestication pathways, domestication syndrome, insect species

1. Introduction

Domestication is one of the most important developments in human history [1]. Beginning during the Late Pleistocene with dog domestication [2, 3], it has irrevocably impacted human history, demography, and evolution leading to our current civilizations [1, 4–6]. Domesticated species play important roles for humans in many aspects of our daily life by providing food, biological control agents, pets, sporting animals, basic materials, and laboratory models [1, 7, 8]. This considerable importance in our culture, survival, and way of life has always aroused the curiosity of scientists and nonscientists. An extraction from the database Scopus of articles and reviews published since 1960 in Life Science Area (i.e., agricultural and biological sciences; biochemistry, genetics and molecular biology; environmental science; multidisciplinary) for which the term "domestication" is cited in the title, the abstract, or the keywords inventories 6199 documents (database accessed on August 31, 2018). However, despite this profusion of literature, significant questions regarding the domestication process, the domesticated species notion, or the domestication histories still remain [9–11].

The notions of domesticated species and domestication process are among the most confusing and controversial concepts in biology [12–14]. Vivid debates are continually fuelled by clashes of conflicting, although complementary, visions of botanists, mammalogists, ornithologists, ichthyologists, archeologists, geneticists, and sociologists. The achievement of a consensual view is impeded by the complexity of the domestication phenomenon, which involves many phylogenetically distant species and occurs in several different social and cultural contexts [1]. Nevertheless, there were some attempts to unify the alternative points of view to some extent [1, 12, 13, 15–17]. For the purpose of this chapter, domestication can, thereby, be considered as the process in which populations are bred in man-controlled environment and modified across succeeding generations from their wild ancestors in ways making them more useful to humans who control, increasingly during the process, their reproduction and food supply [1, 12, 15–17]. This process does not involve all populations of a particular species: some populations can undergo domestication, while other populations do not. The domestication process is a continuum that can be divided into five key steps (the so-called "domestication levels") based on the degree of human control over the population life cycle and the degree of gene flow from wild counterparts [12]. This classification had been primarily developed for fish species [12, 18] but can be extended to other species (**Figure 1**). At the early stage (level 1) of the domestication process, the first attempts of acclimatization of a wild population to man-controlled environments are made [12]. These environments can be captive or "ranch" conditions quite isolated from wild populations where living conditions, diet, and food are controlled by humans [19]. The next stages correspond to an increasing control of the life cycle by humans: level 2—a part of life cycle is controlled by humans in man-controlled environments, but "seed" materials are collected in the wild to maintain rearing of the species (i.e., capture-based production; e.g., [20]); level 3—the life cycle is fully controlled by humans in man-controlled environments, but significant gene flow from the wild still occurs due to spontaneous introgressions or intentional wild specimen introductions by breeders [21]; level 4—the life cycle is fully controlled by humans in man-controlled environments without wild inputs [12]. The last stage (level 5) corresponds to the development of selective breeding programs or organism engineering to intentionally modify some traits of the human-controlled populations (e.g., [22–24]). Seen from this perspective, a species can be considered as domesticated when it reaches, along this continuum, a threshold arbitrarily defined according to a particular scientific or legislative context. The resulting subjective definition of domesticated species is thus eluded from this chapter.

The domestication process is set during a temporal succession of interactions between a species and humans: the so-called "domestication pathways" [10, 25]. An overview of published domestication histories allows identifying three main pathways [10, 15, 25, 26]. In the commensal pathway, there is no intentional action on the part of humans but, as people manipulated their

http://dx.doi.org/10.5772/intechopen.81834

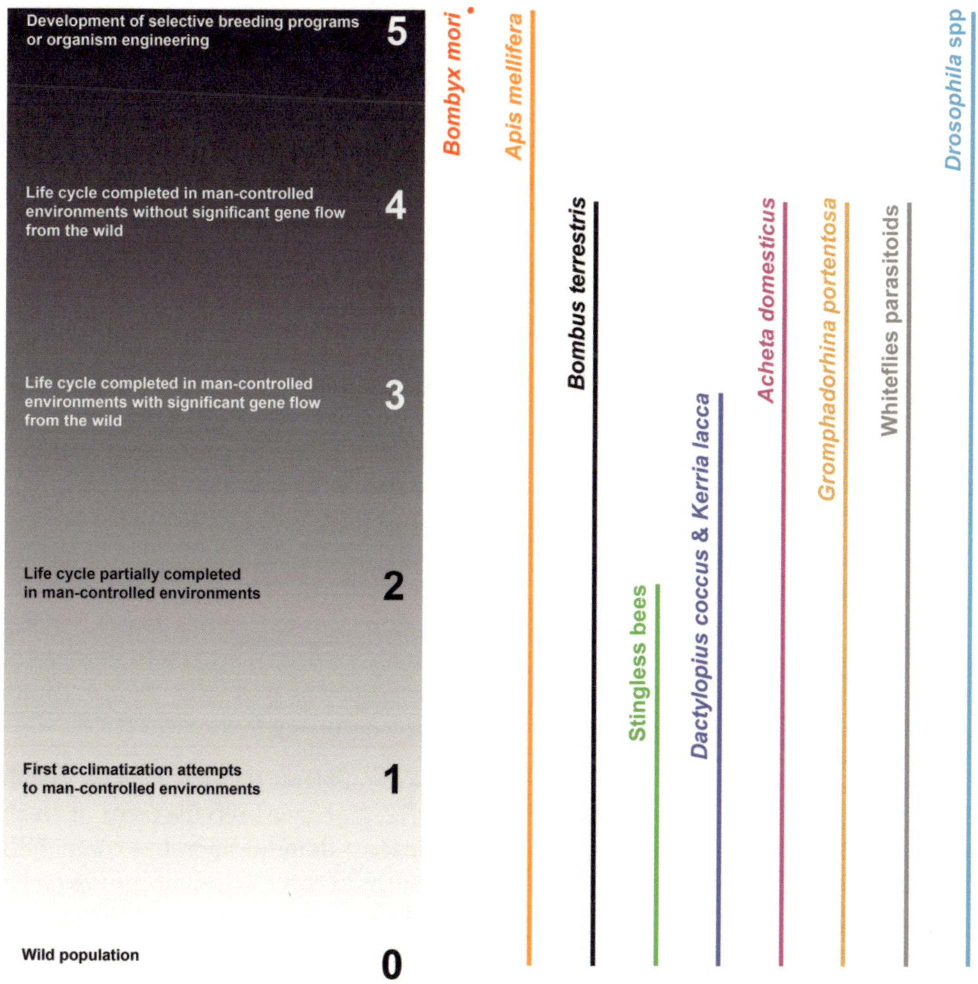

Figure 1. Domestication process and insect domestication level. Numbers 0–5 refer to the domestication levels [12, 18]. Characteristics of each domestication level are provided on the left. Lines and points near the insect species names show the range of domestication degrees observed among populations of the species.

immediate surroundings, some populations of wild species have been attracted to elements of the human niche. The tamer, less aggressive individuals with shorter fight or flight distances of a wild species establish a profitable commensal relationship with humans. Later, succeeding generations of such individuals shift from cynanthropy to domestication through captivity setting up and human-controlled breeding. The dog and the cat are the archetypal commensal pathway species [10]. Contrary to the former, the prey pathway begins with human actions, but the primary human motive is not to domesticate but to increase food resources. Actually, it is initiated when humans modify their hunting strategies into game-management strategies to increase prey availability, perhaps as a response to localized pressure on the supply of prey. Over time and with the more responsive populations (e.g., the more docile individuals), these game-management/keeping strategies turn into herd-management strategies based on a sustained multigenerational control over movements, feeding, and reproduction of populations corresponding to a domestication process. Species that have followed this prey pathway are, for instance, large terrestrial herbivorous mammals [26]. At last, the directed pathway is the only one that begins with a deliberate and directed process initiated by humans in order to domesticate populations of a wild species [26]. Most modern domestic species such as pets [27], transport animals [10], and aquatic species [12, 28] have arisen because of this pathway [10]. The three pathways are theoretical conceptualizations of domestication process, but many species have a more complex history involving several pathways (e.g., pigs [10, 25, 29]).

When the domestication process begins, it results in long-term genetic differentiation and, finally, in the evolution of distinct changes in phenotypic traits [16, 30]. The differentiation of populations undergoing a domestication process can be initiated early in their domestication history and despite persistent gene flow from wild populations [21, 31–34]. The resulting specific morphology, physiology, and behavior constitute the "domestication syndrome" that tends to be more of less similar among different species of a particular organism group [35–40]. Overall, these specificities include domestication traits (i.e., facilitating the early stage of domestication) and improvement traits (i.e., appearing at latter stages of domestication) [35]. The first are shared by all domesticates and generally fixed during the first stages of domestication, while the latter are observed in some domesticated populations when higher human impacts on breeding happens [10]. These changes are driven by (i) selection pressures created by both unintentional and deliberate human actions as well as by human-modified environments and/or by (ii) a relaxation of the selection occurring in the wild [10, 41, 42].

The domestication process, pathways, and consequences on plants (e.g., [1, 37, 43]), mammals (e.g., [1, 10, 26]), birds (e.g., [44, 45]), and fishes (e.g., [12, 28]) have been the focus of an abundant research from Darwin's works [46]. However, insects' domestication has comparatively received far less attention to date [47]. Yet, insects are the most common animal group on Earth: they make up about 75% of all animal species [48, 49]. They play an important role in pollination, waste bioconversion, biocontrol, raw material supplying, food production, medical application, and human cultures. Strangely, major reviews on domestication give the impression that so few have been domesticated [10, 11, 15, 25, 26]. An overview of current literature shows how insect domestication has been overlooked: the database Scopus inventories only 68 papers that focus on it and most of them on only two species (i.e., the silkworm and the honey bee). Actually, most insect rearing/breeding/farming histories have not been considered as domestication processes although they can be interpreted as such. Therefore, the aims of this chapter are (i) to provide an overview of main ancient and recent insect domestication histories and (ii) to reread them by the light of the domestication process, pathways, triggers, and consequences observed in other animal species. Some of the considered species (silkworm and honey bee) have been chosen because they are among the few insects commonly acknowledged as domesticated species, while others have been considered since they allow illustrating alternative domestication patterns.

2. The silkworm and the sericulture

Silkworm is the caterpillar of the moth *Bombyx mori* (Lepidoptera, Bombycidae). It is one of the most important insects in human economy because the species is the primary producer of silk

[50, 51]. Although silk has a tiny percentage of the global textile fiber market (i.e., less than 0.2%; the yearly worldwide production is about 200,000 metric tons of silk [51]), the annual turnover of the China National Silk Import and Export Corporation alone is more than 2 billion US$ [19, 51]. Moreover, silk production provides employments to several million persons in rural and semirural areas across the world [19] (e.g., 8 millions in India [51]). Beside its economic importance, *B. mori* is an edible insect [19], a health food [19], a pet [19], and model species for basic research because of its short life cycle and adaptation to laboratory culture [52–55].

2.1. *Bombyx mori* life cycle and production

The silkworm life cycle is strongly controlled by humans in indoor facilities with controlled environmental conditions [51]. New eggs are incubated in rearing facilities where their hatching can be scheduled and synchronized by humans through chemical treatments and photothermal controls (e.g., black boxing practices) [51]. The newly hatched caterpillars are transferred to rearing tray (i.e., brushing process) and fed by humans with man-produced plants (e.g., mulberry leaves) [51]. After several molts, caterpillars climb on man-provided supports and spin their silken cocoons. Then, cocoons are collected and *B. mori* specimens are killed before metamorphosis since proteolytic enzymes released to make a hole in the cocoon by the adults are destructive to the silk [51]. Some cocoons are allowed to survive in order to produce adults for breeding [51]. In contrast to closely related wild moth species (e.g., *B. mandarina*) that fly for reproduction or evasion from predators, *B. mori* adults are not capable of functional flight due to their too big/heavy body and their small wings [51]. Therefore, *B. mori* completely relies on human assistance in finding a mate and a laying support [51]. The *B. mori* oviposition site selection is also controlled by humans (i.e., egg laying occurs on man-offered mulberry plant or on filter paper) [51, 56].

2.2. Domestication history and pathway of *Bombyx mori*

Bombyx mori is one of the few insects commonly acknowledged as truly domesticated and as a stunning case in point of insect domestication [47, 52, 57, 58]. Several archeological and molecular studies have tried to trace the history of its domestication (e.g., [57, 59–62]). The silkworm was domesticated roughly 7500 years ago from Chinese populations of *B. mandarina*, an extant wild silk moth of East Asia [57, 59, 60, 63]. The domestication of the silkworm is thought to be a directed pathway [10] starting at a single event [61]. Long-term bidirectional significant gene flow occurred between wild and domesticated silkworm populations during the first 3500 years of the domestication [59] most likely because of accidental escapes and intentional hybridizations by breeders to produce desirable strains [52, 59, 64]. Nowadays, low gene flow presumably still exists with *B. mandarina* [65].

Even though silk spread rapidly across Eurasia, its production remained exclusively Chinese for several millennia [62, 66]. Indeed, the sericulture (i.e., the raising silkworms for silk production) spread only to Korea and Japan around 2000 years ago [57, 60] and was even later introduced to Central Asia and Europe (i.e., the Byzantines acquired the sericulture methods by 522 CE) through the Silk Road [57, 66]. This silkworm production expansion is one of the most tremendous examples of the direct and indirect consequences of the animal domestication on the human history [57]. Indeed, the opening of Silk Road has dramatically impacted human history by triggering cultural/technical/good exchanges as well as population movements and disease spread out (e.g., bubonic plague) between Eurasian civilizations while its closing forced the merchants to take to the sea to ply their trade triggering the Age of Discovery [51, 66]. The industrial revolution and the increasing demand in Europe led to a peak of the sericulture by the eighteenth and nineteenth centuries before declining due to silkworm disease breakouts and the raising of cotton industry [51].

2.3. Consequences and progress of the domestication process in *Bombyx mori*

Bombyx mori displays significant specificities compared to its phylogenetically nearest wild counterpart [67–71]. Some of these traits can be considered as (i) domestication traits reinforced by or (ii) improvement traits fostered by selective pressures shaped by unintentional/deliberate human actions and human-modified environments: an increased cocoon size, larger body size, higher silk production, higher growth rate, larger tolerance to human presence/handling, higher ability to live in crowded conditions, and a better feed efficiency [51, 52, 57]. Conversely, other specificities could be explained by a relaxation of the selection occurring in the wild (e.g., predation pressure): leucism (meaning the loss of camouflage) and disability to fly [51, 68]. These last changes have made *B. mori* entirely dependent upon humans for survival, feeding, and reproduction [51, 52]. Moreover, independent selective breeding programs and different breeding environments (i.e., from temperate to tropical climate) have led to the development of more than 1000 inbred lines or strains of domesticated silkworms across the world [51, 57, 60, 72]. Since *B. mori* (i) has its life cycle fully controlled by humans in captivity, (ii) is entirely dependent on humans for reproduction, (ii) and undergoes selective breeding and genetic improvement to harvest maximum output, they are one of the few insect species at a very advanced domestication stage (Level 5; **Figure 1**). While they are not as extreme as the *B. mori* case, other moth species used for silk production have their life cycle under human control and dependence such as *Samia cynthia* (i.e., ericulture; see [73, 74]).

3. The honey bees: beekeeping or apiculture?

Honey bees are eusocial insect species distinguished by their production and storage of honey and their construction of colonial nests from wax [75]. They belong to the same genus (Hymenoptera, Apidae, *Apis* spp.) that includes 11 species and many subspecies native from the Old World [75, 76]. The dwarf honey bees (*A. florea* and *A. andreniformis*) are small species from southern and southeastern Asia that make small open nests in trees and shrubs [75, 77, 78]. These species produce honey that is harvested and eaten by local human populations [77, 79]. The giant honey bees (*A. binghami, A. breviligula, A. dorsata,* and *A. laboriosa*) are aggressive species inhabiting forest areas of South and Southeast Asia [80–82]. They produce honey and wax in their open nest on trees, cliffs, or buildings that are harvested by indigenous people [83–85]. *Apis koschevnikovi* and *A.nuluensis* are cavity-nesting species that occur in the tropical evergreen forests of Borneo [86, 87]. *Apis nigrocincta* is a cavity-nesting species reported in Sulawesi [75]. The western honey bee (*A. mellifera*) and the eastern honey bee (*A. cerana*) are cavity-nesting species native throughout (i) Africa, the Middle East, and Europe and (ii) South and Southeast Asia, respectively [75]. All *Apis* species are important pollinators for many ecosystems [88].

Although other species like *A. dorsata* or *A. cerana* can be important for human economy and feeding in certain countries, none achieves the crucial economic, agricultural, scientific, and environmental importance of *A. mellifera* [89–91]. Its importance relies on its pollination activity as well as on its production of honey, wax, venom, pollen pellets, propolis, and royal jelly [92].

3.1. *Apis mellifera* life cycle and production

Unlike most of other bee species, honey bees produce perennial colonies with large number of individuals that (i) belong to different castes (i.e., workers that are sterile females, drones that are males, and queen that is the reproductive female) and (ii) are not able to survive by themselves for extended periods [75]. In the nest, there is a labor division between castes: (i) the workers harvest pollen and nectar on flowers to feed larvae, queen, and other workers as well as to store food as honey [89, 93] and protect the nest from predators and (ii) queen ensures the production of new queens, drones, and workers [75]. The colony is considered as a superorganism since it is a collection of agents, which can act in concert to produce phenomena (e.g., colony exhibit homeostasis and emergent behavior) governed by the collective [94]. When environmental conditions are favorable (i.e., abundance of food), new queens are produced while old queen with up to two-thirds of the workers leaves the nest in a swarm to find a new location to establish a new nest [89]. In the old nest, new queens compete until only one remains and the survivor takes the nest control [89]. Then, the new queen goes on one or more nuptial flights and mates with several drones [95]. Once mating is done, the queen remains in the hive and lays eggs [89]. The swarming behavior and the takeover of the old nest by the new queen can be interpreted as the reproduction of the superorganism.

Humans can control the life cycle of the superorganism by providing man-made hives for the colony to live and store food [89]. This allows humans to easily collect honey and other products that hive produces rather than to scavenge these products in the wild. More advanced practices allow apiarists to control colony reproduction by restricting swarming behavior and controlling mating by artificial insemination [96, 97].

3.2. Domestication history, traits, and pathway of *Apis mellifera*

Molecular dating suggests that *A. mellifera* expanded its distribution around 1 million years ago [98, 99] from a still debated ancestral range [76, 90, 98–102]. During its range expansion, the western honey bee experienced local adaptations [103] and geographic differentiations leading to the current substantial phenotypic variation across its extensive geographic range [101]. This intraspecific variability has been used to develop an extensive classification of 29 subspecies (or "races") [76]. These taxa are now lumped into four major groups based on morphological, genetical, ecological, physiological, and behavioral traits: the African, Western/Northern European, Eastern European, and Middle East populations (review in [100]). The European groups exhibit phenotypic adaptations to survive colder winters, whereas the African group is more aggressive and shows a greater tendency to swarm [101].

Humans began harvesting wax and honey from honey bee colonies at least 9000 years ago [104, 105]. They originally scavenged these products from wild nests [89, 104, 105]. However, the demand for honey outgrew its natural availability as human populations became larger and sedentary [106]. This context presumably triggered the beekeeping development by providing hives to honey bees that make it easier to harvest their honey and wax by humans [105]. At the beginnings of beekeeping, honey bees were not "bred" so much as "kept": humans provided rudimentary containers (often destroyed during honey harvesting) and hoped that wild bee colonies would take up residence without later swarming [105]. Over time, humans increased their control on bees by developing swarming control device (i.e., queen excluder [96]), reproduction control (e.g., artificial insemination [97]), mass breeding (e.g., [107]), selective breeding programs (e.g., [108–110]), and new strains (e.g., Buckfast strain [111] or Africanized honey bees [112]).

The honey bees' domestication concerns only *A. mellifera* and *A. cerana* (see details about the later species in [89]) most likely because they display intrinsic features that facilitated the domestication process: (i) cavity-nesting habit making hives suitable for these species, (ii) hygienic behavior (i.e., detection and removal of diseased brood and wastes) limiting diseases, and (iii) adaptations to tropical and temperate climate facilitating the apiculture development across the world [89, 110], for example, *A. mellifera*. Moreover, differentiations in traits facilitating beekeeping are observed at the subspecies level. Subsequently, some particular subspecies were preferably domesticated by humans. For instance, non-African subspecies have been more widely used by most beekeepers since they can survive in temperate regions, have a low tendency to swarm, and low aggressiveness [101].

Domestication history of honey bees has been investigated through molecular datasets that highlight several domestication events followed by introgression between subspecies [90, 113, 114]. Although the honey bee domestication history has been regarded as a directed pathway [10], the evolution from early beekeeping practices to modern apiculture practices can been seen as similar to the prey pathway in which game-keeping strategies turns into control over movements, feeding, and reproduction. However, it is likely than directed and prey pathways occurred during honey bee domestication history since several domestication events happened [90, 113, 114].

3.3. Is *Apis mellifera* domesticated?

Many authors acknowledge (often without justification) the domesticated status of *A. mellifera* (e.g., [10, 16, 47, 58, 89, 102, 115–117]). In contrast, *A. mellifera* has been considered as never properly domesticated but only as managed species by other authors (e.g., [110, 114]; however, some of these scientists acknowledge an ongoing domestication process) because (i) their biology, physiology, and behavior are seen as largely unchanged from their wild counterparts [114], (ii) honey bees are able to survive without human's help [118], (iii) there is extensive gene flow between wild/feral and managed bees in native range due to the difficulties to achieve controlled mating [119]. However, these points should be reconsidered. First, the comparison of phenotypes between "wild" and "nonwild" populations is difficult in a large a part of the species range. Indeed, colonies that are found in the wild may have escaped from a managed colony, and therefore, they may not be wild [120]. In Europe, it is unlikely that there are any truly wild subpopulations left due to this gene flow [120]. This means that the differentiation fostered by the domestication process can be blurred by the large amount of feral populations in the wild. Nevertheless, there are significant behavioral changes observed in man-controlled honey bees stocks such as multiple queen colonies (i.e., colonies conserved several queens without deadly competition between them [121]), decreased aggressiveness, higher honey production, increased foraging zeal, and disinclination to swarm of some strains [111]. These specificities can be interpreted as improvement traits within a domestication syndrome. Second, many other species acknowledged as "domesticated" can survive in the wild (e.g., feral populations of rabbits, cats, and dogs [122]; although fast initial decline in fitness of domesticated escapees in the wild is expected [123]). Moreover, the ability of honey bees to survive in the wild could be overestimated since most *A. mellifera* are not considered to be self-sustaining as veterinary treatments against the mite *Varroa destructor* among other parasites is often provided [120]. Third, gene flow between "nonwild" and wild populations is commonly observed during the domestication process (see [21, 124, 125]). Actually, the debate about the status of domesticated animal for *A. mellifera* exemplifies the subjectivity of the domestic species threshold. Beside this controversial definition, *A. mellifera* shows that different conspecific populations can be at different stages of the domestication process. Indeed, there is no control by humans over the life cycle of wild populations that are commonly observed for the African group [126–128]. In contrast, many populations belonging to the European groups have a life cycle completed in man-made environment (i.e., hives) and controlled by humans (i.e., control of superorganism reproduction), feed on domesticated crops (i.e., humans can actively control the honey bee food supply for honey production or crop pollination) and/or on artificial food provided by humans (i.e., sugar syrup) [129], and some of them undergo selective breeding programs [108–111]. Therefore, the domestication levels of *A. mellifera* range from 0 to 5 according to the population considered.

4. The bumble bees and the stingless bees: the other bee domestications

About 90% of world's plant species are pollinated by animals [130–132], and the main animal pollinators in most ecosystems are bees [88]. Although other taxa like butterflies, flies, beetles, wasps, or vertebrates can be important pollinators in certain habitats or for particular plants [133, 134], none achieves the numerical dominance as flower visitors worldwide as bees [130, 131]. The pollination efficiency of bees has been used by humans to improve their crop yields. The western honey bees is the most commonly used species in managed pollination service [76, 135]. This species pollinates nearly half of the top 115 global food commodities and is capable of increasing the yields of 96% of animal-pollinated crops [117, 136]. However, the lack of sufficient stocks of honey bees to ensure pollination service [115, 137], the aggressiveness of Africanized honey bees (i.e., obtained by man-made hybridization between African and European subspecies of *A. mellifera* to breed a strain of bees that would produce more honey and be better adapted to tropical conditions) in Neotropics [138], and the poor pollination efficiency of *A. mellifera* for some plants, as well as the requirement of maintaining the honey bee colonies outside the flowering period of valuable crops [139] have triggered or restarted the domestication of other bee species: the bumble bees and the stingless bees.

4.1. The bumble bees

Bumble bees (Hymenoptera, Apidae, *Bombus* spp.) are social insects with a nearly worldwide distribution with their largest species diversity in temperate and cold areas [75, 140]. Except in tropical regions, bumble bees produce annual colonies (i.e., colony dies after the production of new queens and males). They have several adaptations such as their ability to "buzz pollinate" (i.e., sonication) and their insulated bodies that make them ideal pollinators for many valuable crops (e.g., raspberry and tomatoes) for which honey bees are quite inefficient pollinators [139–141]. Moreover, bumble bee colonies can be easily shipped and can be maintained without specialists' help in crops fields [139]. However, they do not produce honey or hive material suitable for mass market sales [117]. Therefore, their commercial importance is only based on their pollination efficiency on particular crops, mainly in greenhouses [117]. First trials of bombiculture were attempted by researchers at the beginning of the twentieth century [142, 143], but bumble bee mass production started during the 1980s [139, 144]. At least five species have underwent domestication process for pollinator production (*B. ignites*, *B. impatiens*, *B. lucorum*, *B. occidentalis*, and *B. terrestris*) and many more for research purpose [139, 142–145]. Their domestication history follows a directed pathway with several domestication events since several bumble bee breeders started independently the production of the same taxon [139, 144]. *Bombus terrestris* is the most traded species: more than 2 million colonies are yearly produced and shipped throughout the world (review in [146]). The species is mass produced in indoor facilities in which the whole life cycle is controlled by humans [139]. The choice of the most effective bumble bee species for mass production has been made through the test of several species [139]. Most of them have been proved to be poorer choices (e.g., more aggressive, hard to feed, low production success), and their domestication programs have been abandoned [139, 144]. This means that the domestication process of such species (e.g., *B. lapidarius* [139]) has regressed after a period of human interest.

Although domestication of bumble bees has been acknowledged by various authors (e.g., [139, 147]), comparison between breeders' stocks and wild populations is still lacking to highlight potential domestication syndrome in *Bombus* species. Nevertheless, the domestication programs of the five most produced species are already quite advanced (Level 4, **Figure 1**) since current bumble bee breeders' stocks experience low gene flow with their wild counterparts. However, no selective breeding program has been reported to date.

4.2. The stingless bees

Stingless bees (Hymenoptera, Apidae, Meliponini) are social bees with perennial colonies (i.e., nest can remain active for more than 50 years) occurring in most tropical or subtropical areas [75, 148]. They are known for their pollen/honey production and their pollination efficiency for several valuable crops (e.g., coffee, Avocado, Strawberry, Rambutan) [138, 148]. Meliponiculture dates back to the Maya civilization and is nowadays practiced in Australia and Central/South America [148–150]. Nevertheless, their domestication process has not progressed so far (Level 2, **Figure 1**) since most of the meliponiculture is mainly a capture production that consists in attracting stingless bee swarms and maintaining the colonies in artificial wooden hives [148, 150].

5. Cochineal insects

Scale insects (Hemiptera, Coccoidea) are the third large insect groups including species that are, sometimes, considered as domesticated [47, 58, 151]: cochineals, lac scales, *Cerplastes* species, and *Ericerus* species. These species are economically valuable for the substances that they produce under proper farming [152]. *Dactylopius coccus* and *Kerria lacca* are among the most produced scale insects.

5.1. *Dactylopius coccus*

Cochineal is an important source of red for dyes, lake pigments, cosmetics, and food/pharmaceutical colorants [151, 153]. Indeed, the red dye is mainly composed of carmine, which is a pigment obtained from the scale insects belonging to *Dactylopius* genus (Hemiptera, Dactyloidae) or some *Porphyrophora* species (Hemiptera, Margarodidae) [151, 153–155]. Nowadays, most of the carmine production is based on the farming of *D. coccus* inhabiting (sub)tropical South and Central America [151, 153].

The species is used as a source of carmine in Mesoamerica and South America since the pre-Columbian times [156]. The earliest known cochineal-dyed textiles dates back to the twelfth century, but first evidence of cochineal farming is estimated to the tenth century [155–157]. The center of domestication is thought to be in Mexico [157]. Carmine became an important export good during the Spanish colonial period [156]. Later, the species was introduced in other areas such as Australia, Canary Islands, South Africa, and South Asia [156]. In the middle of the nineteenth century, the production of cochineal fell sharply due to the development of artificial red dyes. Consequently, the cochineal trade almost totally disappeared in the twentieth century. Since the 1970s, cochineal production was restarted due to the discovery of carcinogenic and hazardous properties of synthesized dyes [155].

Dactylopius coccus females are wingless sessile parasites of cacti in the genus *Opuntia* [151, 153, 155]. After mating with winged males, females give birth to nymph that are airborne transported by the wind to new host plants. Since the *D. coccus* females depend on *Opuntia* plants, their production takes place in cactus farms producing domesticated plant species [156, 158]. Cochineals are produced by infesting cacti plants and harvesting the insects by hand 90 days later. During this period, humans actively control potential predators [159]. At the end of the process, some cochineals are left to reproduce, while others are collected for carmine extraction.

Dactylopius coccus has been considered as a domesticated species [47, 58, 151], since it is reliant on human propagation and protection for survival at least in some regions [159]. Moreover, most of *D. coccus* is produced in man-controlled environments (i.e., cactus farms). Nevertheless, humans poorly control the species reproduction and gene flow with wild populations is frequent; yet more recent developments are improving this control (e.g., environment-control microtunnels in Mexico [58]). Overall, the domestication process of *D. coccus* (level 3, **Figure 1**) is far behind the ones of silkworms and honey bees. Since comparison between farmed and wild cochineals is lacking, potential domestication syndrome is unknown.

5.2. *Kerria lacca*

Lac is an important commercial resin of several utilities (e.g., material construction, cosmetics, medicine). It is a resinous secretion of lac insect species from Asia and Central America

[160, 161]. *Kerria lacca* (Hemiptera, Kerriidae) is one of the main species used for lac production [160, 161]. Its life cycle is similar to *D. coccus* with winged males and wingless sessile females that parasite several hundred host plants [161, 162]. For several centuries, lac yields were collected from the wild on infested host plants by local human populations [161]. During the nineteenth century, the increase of exportation from Asia triggered the development of artificial inoculation and mass production [161] through a domestication history that can be interpreted as a prey pathway (i.e., human control on the species was triggered by the need of increasing lac supply). Similarly to *D. coccus*, the domestication process of *K. lacca* is at an early stage (level 3, **Figure 1**) since the current production involved only host plant, lac crop, and lac pest management.

6. Farmed edible and medicinal insects

Humans have been eating insects for millennia [58, 163]. However, human entomophagy is a long-standing taboo in westernized societies [19, 58, 164]. This can explain why insect farming for human food supply has been largely absent from the main agricultural innovations and domestications with few exceptions such as honey bees, silkworms (i.e., pupae is a by-product of silk production), and scale insects [19, 73]. Yet, more than 2 billion of people eat insect regularly since there are a source of protein, fat, vitamins, and minerals frequently stored and sold in developing countries (review in [73, 164]). Across the world, more than 2000 insect species are considered as edible for human food or animal feed [19, 58, 164, 165]. Beside food, insects provide many natural products for drugs to treat human diseases [166, 167].

Overall, the most commonly consumed insects by humans or livestock/pets are beetles (Coleoptera) (31%), caterpillars (Lepidoptera) (18%), bees/wasps/ants (Hymenoptera) (14%) as well as crickets (Orthoptera) (13%) [19, 58, 73, 163–165]. Most of these insects, as well as those used as entomoceuticals, are harvested in the wild [163] but some of these species are farmed for sale and profit [19, 73]. Currently, commercially farmed insects include (i) the house cricket (*Acheta domesticus*), the palm weevil (*Rhynchophorus ferrugineus*), the giant water bug (*Lethocerus indicus*), and water beetles (various species of Coleoptera) for human consumption [58, 168, 169] and (ii) bees, wasps, flies, butterflies, moths, and cockroaches for drug production [167]. Even in small-scale production in developing countries [19], their production implies that their life cycle is controlled by human in captive conditions isolated from their wild counterparts in order to meet regulations about human food production (i.e., hygienic standards, sterile conditions) as well as limiting pathogen spillover from/to the wild [19, 164, 169–171]. Such conditions are conductive for an advanced domestication process (Level 4, **Figure 1**) through a directed pathway. Conversely, other species are produced through an increasing human manipulation of their environment to increase insect yields and to ensure their long-term availability as food [172]. For instance, edible social wasps (Hymenoptera, Vespidae, *Vespula flaviceps,* and *V. shidai* in Japan) are traditionally managed by keeping wasp nests collected in the wild in hive boxes during one season to improve yields [173]. However, current attempts to improve the practice involves efforts to maintain new queens in captive condition over several generations [173], paving the ways to a prey domestication pathway.

7. Biological control agents and sterile insect technique

Addressing the needs of the increasing human population will require a 60% increase in global food production by 2050 [174]. Insects could aid in achieving this objective by providing food production [19, 164] as well as pollination service (see Section 4) and biological control of pests [175].

Biological control is a method of controlling pests such as arthropods, weeds, and plant diseases using predator (e.g., ladybugs to control aphids [176], herbivorous, or parasite species [175]). Parasitoids are among the most widely used biological control agents (e.g., [177, 178]). In these species, female deposits its egg inside or outside a host where emerged parasitoid larva continues to feed resulting in the host death [178–180]. This parasitic way of life is used by humans to target hosts that are pests. Whiteflies parasitoids (Hymenoptera, Aphelinidae, Encyrtidae, Eulophidae, Platygastridae, Pteromalidae, and Signiphoridae) are an example of insects used in greenhouses to control major crop pests (i.e., the whiteflies; Hemiptera: Aleyrodidae) [177, 180]. As many other parasitoids (e.g., fly *Eucelatoria*, the beetle *Chrysolina*, and the wasp *Aphytis*), they are massively produced in captive conditions by humans before being shipped across the world [180]. The full control of their life cycle by humans is needed in order to ensure that the production (i) matches with the appropriate release dates when susceptible host species is at a suitable phase of development [181] and (ii) is available on a yearlong basis to response to demand across the world [178, 182].

The sterile insect technique (SIT) is an alternative approach to control main pests (e.g., [183–185]) or disease vectors (e.g., [186–188]). This method implies to massively release sterile males (sterilized through the effects of irradiation on the reproductive cells) of an insect species into a target environment to compete with wild males for reproduction [183–185]. Ultimately, mass releases allow limiting offspring production of a particular pest and promoting its eradication (e.g., [184]). Mass-rearing production with a life cycle fully controlled by humans is needed to produce the large quantity of insect required by SIT [183].

The required full control of life cycle of pest insects for SIT or biological control agents means that an advanced domestication process is reached (up to 5 since some patented strains are available [189]). In the context of SIT, several studies have investigated the differences between wild and mass-produced males in order to ensure that released sterile males are able to compete with wild males (e.g., [183, 190]). These studies show that the domestication process has triggered several ecological and behavioral divergences between produced and wild populations as well as a decreased fitness of produced populations in the wild (e.g., [183, 190]).

8. Insects as pets

Archeological pieces of evidence show that insects have been used as pets for centuries [191]. Nowadays, crickets, grasshoppers, beetles, cockroaches, silkworms, ants, honey bees, bumble bees, mantises, and stick/leaf insects are bred by humans as a pleasing activity or for teaching purpose [192–194]. Conversely to vertebrates [8, 195–197], there is no, to my knowledge, scientific literature addressing the domestication of pet insects. However, some of these pet insects are produced for other purpose such as honey bees, silkworms, and house crickets for which a domestication process is acknowledged (see previous sections). For other species, such as hissing cockroach (*Gromphadorhina portentosa*), mass/small-scale, and/or amateur production are practiced [198–202]. As for other "exotic" pets (e.g., [18]), these productions involve (i) a full control by humans on the life cycle in captive conditions since a large part of the production is completed out of the species native range and (ii), thus, an advanced domestication process (level 4, **Figure 1**).

9. Insects for laboratory research

Animals are widely used as model species in biology and biomedical sciences. Some insect species have been used for laboratory experiments for several decades (e.g., silkworms, honey bees, and other species [54, 203, 204]), especially the fruit flies (*Drosophila* spp.) [205–207].

Drosophila species first entered laboratories about 1900 and are now standard laboratory animals [208, 209]. As they become an instrument for scientific production, *Drosophila* have been massively produced in laboratory conditions in which life cycle, feeding, and mating are highly controlled by humans [208, 210–212]. This human control along with the strain development and artificial selection for particular purposes [208, 213–216] reflect an advanced domestication process of some populations (level 5, **Figure 1**), while there are many wild populations (e.g., [206, 217, 218]).

Conversely to most other insect species, domestication of *Drosophila* populations has been the focus of several studies since it has been considered as a model system to understand the consequences of the domestication process on genomes and phenotypes [219]. Indeed, fruit flies are easy and cheaply to bred and have a rapid generation time (i.e., at least a dozen generations per year) [206, 220]. This allows comparing several populations that have or not been subject to different domestication histories (e.g., [221–223]) or even monitoring evolutionary trajectories of population undergoing a domestication process since their foundation from the wild [219, 224–226]. This has allowed studying domestication process in well-defined laboratory experiments with replication and specific environmental controls for several *Drosophila* species. An overview of these experiments allows highlighting the domestication consequences for *Drosophila* taxa. Different studies highlight that "domesticated" populations display genetic specificity and accumulation of deleterious mutations, inbreeding depression as well as increasing of fertility, tameness, and manageability due to selection for human-accommodating phenotypes and/or the relaxation of selection on traits adapted in nature [219, 220, 222, 227–230]. Moreover, the evolutionary convergence is observed between long-established laboratory populations [219, 220, 222, 227–230].

10. Conclusions

10.1. Are insect species undergoing domestication processes?

Although few stunning cases (e.g., *B. mori*) have been the focus of abundant research, scientific literature has poorly investigated insect domestication to date. The main reason of this is that insect domestication for human food supply has been largely absent from the agricultural development with few exceptions [19, 73]. Moreover, it is likely that insect domestication study has been hindered by the complexity and the subjectivity of the definition of domesticated species (e.g., for *A. mellifera* [10, 16, 47, 117–119, 58, 89, 102, 110, 114–116]). The difficulty of defining a threshold along a continuous process is a common problem in biology (see similar debate about the status and the process for the species status *versus* speciation in [231–233]). Consequently, the study of the process is often set aside or eluded due to debates on a particular threshold. In insects, many scientific articles or books (e.g., [234]) have analyzed or reviewed the breeding/productions of various insect species without explicitly describing these processes as domestication. Yet, the human control on the life cycle (i.e., on individuals' life cycle in noneusocial species or on superorganism's life cycle in honey bees) of most produced insect species is congruent with a domestication process (**Figure 1**; *sensu* [12]). Since a large number of insect populations are produced in captive conditions isolated from their wild counterparts (**Figure 1**), many species can be considered as undergoing a domestication process. Moreover, new domestication processes can be expected in the near future due to current challenges to increase human food/sanitary security (e.g., [19, 164, 175, 186–188]) or to address new demands for pets (i.e., similar development to the ornamental fish trade (e.g., [18, 235–237])).

10.2. Domestication patterns in insects

Domestication events in insects are no less complex than in crops and vertebrates. Domestication histories can involve (i) one (e.g., silkworms [61]) or several (e.g., in honey bees and bumble bees [113, 139]) domestication events and (ii) one (e.g., bumble bees [139]) or potentially several domestication pathways (e.g., honey bees). In most insect species (i.e., except for few extreme cases such as silkworms), different populations of a particular taxon can reach different degrees of progress in the domestication process (e.g., from wild status to an advanced domestication level in *B. terrestris*). Gene flow between populations at different domestication degrees is commonly observed in insects [59, 65, 119] but they do not hinder development of domestication syndrome (see next section).

Some insect species undergo domestication processes for several centuries (e.g., *B. mori* and *A. mellifera*; [57, 59, 60, 63, 89, 104, 105]), while domestications of most insects produced as biological control agents, pets, and laboratory organisms, or for SIT strategies and entomoceuticals' production have been recently initiated. These recent domestications have been made possible thanks to the advances in technology of captive environment control and animal food production since the nineteenth century [1]. Indeed, most insect domestications are thought to follow a directed pathway, which requires rapidly a full control of life cycle by humans in man-controlled environments. This implies the use of efficient environment and food control technologies. Technological advances have made possible or easier the domestication of species, which could not be domesticated in the past, paving the way to a new wave of domestication (similarly to aquatic species [28]).

As for vertebrate species (see review in [1, 12]), some intrinsic features can hinder the development of domestication processes: (i) a diet that cannot be easily supplied by humans (e.g., oligolectic bee species feeding only on few plant species), (ii) long life-cycle (e.g., periodical cicadas that spend most of their 13- and 17-year lives underground at larval stage), (iii) bad disposition (e.g., some wasp species), or (iv) reluctance to breed in captivity. Nevertheless, modern technology could potentially allow domesticating any insect species. Indeed, current insect production involves species with very different ecologies (i.e., terrestrial taxa, e.g., silkworm [51]; aquatic species, e.g., water beetles [168]), behavior (i.e., solitary insects, e.g., silkworm [51]; eusocial species, e.g., honey bees [89]), and development (i.e., Endopterygota, e.g., honey bees [89]; Exopterygota, e.g., house crickets [19]); representative of the insect biodiversity. However, new domestication processes, which presumably occur only through directed or prey pathways for insects, are only initiated by humans to provide response to needs or demands of humanity. This means that the domestication of a species that could meet human needs/demands already addressed by another produced species is unlikely [1, 238]. Instead, all species that have recently undergone a domestication process and then have been massively produced are those which provide response to new needs or demands of humanity such as bumble bees (i.e., pollination in greenhouses), hissing cockroach (i.e., pet), or *Drosophila* flies (i.e., laboratory organism) [139, 199, 208, 209].

An overview of current insect productions in man-controlled captive conditions shows that insect taxa are used to address very different human needs (e.g., food [19], raw materials [234], pets [194]). Moreover, many insect taxa that are primary produced to address a specific demand tend to be later used to serve several human needs as observed in the domestication histories of several mammal species. For instance, *A. mellifera* that produces honey (i.e., the primary use) and edible pupae can be considered as the insect equivalent of dairy cows, which are valued not only for their milk but also as meat [19]. Moreover, honey bees provide several raw materials (wax), health food (royal jelly), entomoceuticals (venom), ecosystem service (pollination), model specimens for research [204], and pleasure (recreational beekeeping) to humans [92].

10.3. Domestication consequences and their shaping factors

Overall, differentiations between wild populations and their counterparts undergoing a domestication process have been poorly studied in insect species. Yet, such divergences and convergences of various phenotypic traits that differentiate domesticates from their wild progenitors can be expected under the domestication syndrome hypothesis [36]. In mammals, the domestication syndrome tends to comprise changes in tameness, aggressiveness, coat color/pigmentation, body morphology, reproductive alterations, hormone, neurotransmitter concentrations, and brain composition [36]. Some of these changes can be observed when comparing *B. mori* and its phylogenetically nearest wild counterpart [67–71] in tameness (i.e., larger tolerance to human presence/handling), aggressiveness (i.e., toward conspecifics since *B. mori* has higher ability to live in crowded conditions), morphology (i.e., leucism, larger body size), and reproduction/development (i.e., bigger cocoon and higher silk production, higher growth rate, altered premating behavior) [51, 52, 57, 68]. Comparison of silkworm specificities with phenotypes of man-produced *Drosophila* flies and honey bees shows some convergences: higher tameness (i.e., fruit flies), lower aggressiveness toward humans and conspecifics (i.e., in *A. mellifera*), modified reproduction (e.g., higher fertility in fruit flies; changes in reproduction, e.g., limited swarming in *A. mellifera*), and morphology (i.e., specific color patterns of man-controlled strains/races) [111, 121, 219, 220, 222, 227–230]. These specificities concern domestication traits facilitating the domestication by humans (e.g., aggressiveness in honey bees) as well as improvement traits (e.g., higher honey production in *A. mellifera*; higher silk production in *B. mori*) that increase the manageability and the animal production efficiency/profitability for humans.

Specificities of populations undergoing a domestication process have been most likely shaped by unintentional/deliberate human actions, human-controlled environments, relaxation of the selection occurring in the wild or both as in other animal species [10, 41, 42]. For instance, the inability of *B. mori* to fly could result from a relaxation of selection in the wild (i.e., silkworm are protected and fed in captive conditions by humans) and/or a human pressure for "nonflying" insects (i.e., this facilitates the handling by humans). Similarly, the lower aggressiveness of honey bees can result from a lower predation pressure (i.e., human protection of hives) as well as from human selection for less aggressive populations. Inadvertent human habituation and unintentional conditioning could also be a primary selective agent in insect domestication as suggested to explain developmental and reproductive differences between *Drosophila* strains [220].

From a genetic point of view, animals in captive environment are expected to rapidly display genetic changes corresponding to adaptations to captive breeding [239]. Indeed, the specific selective pressure occurring in domestication environments promotes selection for domestication syndrome gene variants [11]. This selection on man-controlled populations can shape specific genotypes even when gene flow from the wild still occurs [21, 59]. Changes in traits linked to valuable resources for humans or morphology have been showed to have a genetic basis (e.g., specificity of silk gland transcriptomes [67] and melanin synthesis [68] of *B. mori*). Similarly, behavior modifications commonly observed in insect domestication syndrome (e.g., tameness, aggressiveness, manageability by humans) can be explained by mutations on neurogenetic genes affecting overall locomotion and activity as suggested in man-produced populations of *Drosophila* species and mammals [36, 220]. Therefore, large mutational target of neurogenetic genes can explain the evolution of specific behavior in animal populations undergoing domestication processes [220]. These neurogenomic loci collectively provide a large genomic substrate for variation to accumulate, and then selection and drift to act, to transform behavior [220].

10.4. Future prospects

The study of domestication of insect is still at a nascent stage. Some "model species" such as *A. mellifera*, *B. mori*, and *Drosophila* spp. have been the focus of several studies to understand domestication process. However, genetic bases of domestication-fostered modifications as well as the characterization of these modifications are poorly known. Therefore, further studies are needed to generalize domestication patterns as well as to understand genomic basis of domestication process.

Conflict of interest

The author declares no conflict of interest.

Author details

Thomas Lecocq

Address all correspondence to: thomas.lecocq@univ-lorraine.fr

Université de Lorraine, Inra, URAFPA, Nancy, France

References

[1] Diamond J. Evolution, consequences and future of plant and animal domestication. Nature. 2002;**418**:700-707

[2] Larson G, Karlsson EK, Perri A, Webster MT, Ho SYW, Peters J, et al. Rethinking dog domestication by integrating genetics, archeology, and biogeography. Proceedings of the National Academy of Sciences. 2012;**109**:8878-8883

[3] Perri A. A wolf in dog's clothing: Initial dog domestication and Pleistocene wolf variation. Journal of Archaeological Science. 2016;**68**:1-4

[4] Gepts P, Bettinger R, Brush S, Damania A, Famula T, McGuire P, et al. Introduction: The domestication of plants and animals: Ten unanswered questions. In: Gepts P, Bettinger R, Brush S, Damania A, Famula T, McGuire P, et al., editors. Biodiversity in Agriculture—Domestication, Evolution, and Sustainability. Davis, USA: Cambridge University Press; 2012. pp. 1-8

[5] Vigne J-D. The origins of animal domestication and husbandry: A major change in the history of humanity and the biosphere. Comptes Rendus Biologies. 2011;**334**:171-181

[6] Diamond J. Guns, Germs, and Steel: The Fates of Human Societies. New York, NY: Norton; 1997

[7] Kaiser S, Hennessy MB, Sachser N. Domestication affects the structure, development and stability of biobehavioural profiles. Frontiers in Zoology. 2015;**12**:S19

[8] Driscoll CA, Macdonald DW, O'Brien SJ. From wild animals to domestic pets, an evolutionary view of domestication. Proceedings of the National Academy of Sciences. 2009;**106**:9971-9978

[9] Gerbault P, Allaby RG, Boivin N, Rudzinski A, Grimaldi IM, Pires JC, et al. Storytelling and story testing in domestication. Proceedings of the National Academy of Sciences of the United States of America. 2014;**111**:6159-6164

[10] Larson G, Fuller DQ. The evolution of animal domestication. Annual Review of Ecology, Evolution, and Systematics. 2014;**45**:115-136

[11] Larson G, Piperno DR, Allaby RG, Purugganan MD, Andersson L, Arroyo-Kalin M, et al. Current perspectives and the future of domestication studies. Proceedings of the National Academy of Sciences of the United States of America. 2014;**111**:6139-6146

[12] Teletchea F, Fontaine P. Levels of domestication in fish: Implications for the sustainable future of aquaculture. Fish and Fisheries. 2014;**15**:181-195

[13] Bokonyi S. Definitions of animal domestication. In: Clutton-Brock J, editor. The Walking Larder: Patterns of Domestication, Pastoralism, and Predation. London, UK: Unwin Hyman; 1989. pp. 22-27

[14] Cassidy R, Mullin M. Where the Wild Things are Now: Domestication Reconsidered. Berg: Oxford and New York; 2007

[15] Zeder MA. Core questions in domestication research. Proceedings of the National Academy of Sciences of the United States of America. 2015;**112**:3191-3198

[16] Price EO. Behavioral aspects of animal domestication. The Quarterly Review of Biology. 1984;**59**:1-32

[17] Zeder MA. Domestication: Definition and overview. In: Encyclopedia of Global Archaeology. New York, NY: Springer New York; 2014. pp. 2184-2194

[18] Teletchea F. Domestication level of the most popular aquarium fish species: Is the aquarium trade dependent on wild populations? Cybium. 2016;**40**:21-29

[19] Van Huis A, van Itterbeeck J, Klunder H, Mertens E, Halloran A, Muir G, et al. Edible Insects—Future Prospects for Food and Feed Security. Food and Agriculture Organization of the United Nations: Rome, Italy; 2013

[20] Lovatelli A, Holthus PF. Capture-Based Aquaculture—Global Overview. FAO Fisheries Technical Paper: Rome, Italy; 2008

[21] Marshall FB, Dobney K, Denham T, Capriles JM. Evaluating the roles of directed breeding and gene flow in animal domestication. Proceedings of the National Academy of Sciences of the United States of America. 2014;**111**:6153-6158

[22] Vaughan DA, Balázs E, Heslop-Harrison JS. From crop domestication to super-domestication. Annals of Botany. 2007;**100**:893-901

[23] Kerr WA. Selective breeding, heritable characteristics and genetic-based technological change in the Canadian beef cattle industry. Western Journal of Agricultural Economics. 1984;**9**:14-28

[24] Gjedrem T, Robinson N, Rye M. The importance of selective breeding in aquaculture to meet future demands for animal protein: A review. Aquaculture. 2012;**350-353**:117-129

[25] Zeder MA. The domestication of animals. Journal of Anthropological Research. 2012; **68**:161-190

[26] Zeder MA. Pathways to animal domestication. In: Gepts P, Bettinger R, Brush S, Damania A, Famula T, McGuire P, et al., editors. Biodiversity in Agriculture—Domestication, Evolution, and Sustainability. Davis, USA: Cambridge University Press; 2012. pp. 227-259

[27] Fritzsche P, Neumann K, Nasdal K, Gattermann R. Differences in reproductive success between laboratory and wild-derived golden hamsters (*Mesocricetus auratus*) as a consequence of inbreeding. Behavioral Ecology and Sociobiology. 2006;**60**:220-226

[28] Duarte CM, Marba N, Holmer M. Rapid domestication of marine species. Science. 2007;**316**:382-383

[29] Frantz L, Meijaard E, Gongora J, Haile J, Groenen MAM, Larson G. The evolution of Suidae. Annual Review of Animal Biosciences. 2016;**4**:61-85

[30] Jensen P. Domestication—From behaviour to genes and back again. Applied Animal Behaviour Science. 2006;**97**:3-15

[31] Larson G, Albarella U, Dobney K, Rowley-Conwy P, Schibler J, Tresset A, et al. Ancient DNA, pig domestication, and the spread of the Neolithic into Europe. Proceedings of the National Academy of Sciences of the United States of America. 2007;**104**: 15276-15281

[32] Naderi S, Rezaei H-R, Taberlet P, Zundel S, Rafat S-A, Naghash H-R, et al. Large-scale mitochondrial DNA analysis of the domestic goat reveals six haplogroups with high diversity. PLoS One. 2007;**2**:e1012

[33] Ji R, Cui P, Ding F, Geng J, Gao H, Zhang H, et al. Monophyletic origin of domestic bactrian camel (*Camelus bactrianus*) and its evolutionary relationship with the extant wild camel (*Camelus bactrianus ferus*). Animal Genetics. 2009;**40**:377-382

[34] Clotault J, Thuillet A-C, Buiron M, De Mita S, Couderc M, Haussmann BIG, et al. Evolutionary history of pearl millet (*Pennisetum glaucum* [L.] R. Br.) and selection on flowering genes since its domestication. Molecular Biology and Evolution. 2012;**29**:1199-1212

[35] Olsen KM, Wendel JF. A bountiful harvest: Genomic insights into crop domestication phenotypes. Annual Review of Plant Biology. 2013;**64**:47-70

[36] Wilkins AS, Wrangham RW, Fitch WT. The "domestication syndrome" in mammals: A unified explanation based on neural crest cell behavior and genetics. Genetics. 2014;**197**:795-808

[37] Weeden NF. Genetic changes accompanying the domestication of *Pisum sativum*: Is there a common genetic basis to the 'domestication syndrome' for legumes? Annals of Botany. 2007;**100**:1017-1025

[38] Sakuma S, Salomon B, Komatsuda T. The domestication syndrome genes responsible for the major changes in plant form in the Triticeae crops. Plant & Cell Physiology. 2011;**52**:738-749

[39] Santos-Del-Blanco L, Alía R, González-Martínez SC, Sampedro L, Lario F, Climent J. Correlated genetic effects on reproduction define a domestication syndrome in a forest tree. Evolutionary Applications. 2015;**8**:403-410

[40] Lorenzen K, Beveridge MCM, Mangel M. Cultured fish: Integrative biology and management of domestication and interactions with wild fish. Biological Reviews. 2012;**87**:639-660

[41] Mignon-Grasteau S, Boissy A, Bouix J, Faure J-M, Fisher AD, Hinch GN, et al. Genetics of adaptation and domestication in livestock. Livestock Production Science. 2005;**93**:3-14

[42] Zohary D, Tchernov E, Horwitz LK. The role of unconscious selection in the domestication of sheep and goats. Journal of Zoology. 1998;**245**:129-135

[43] Meyer RS, DuVal AE, Jensen HR. Patterns and processes in crop domestication: An historical review and quantitative analysis of 203 global food crops. The New Phytologist. 2012;**196**:29-48

[44] Tixier-Boichard M, Bed'hom B, Rognon X, Bed'hom B, Rognon X. Chicken domestication: From archeology to genomics. Comptes Rendus Biologies. 2011;**334**:197-204

[45] Sawai H, Kim HL, Kuno K, Suzuki S, Gotoh H, Takada M, et al. The origin and genetic variation of domestic chickens with special reference to junglefowls *Gallus g. gallus* and *G. varius*. PLoS One. 2010;**5**:e10639

[46] Darwin C. The Variation of Animals and Plants under Domestication. 1st ed. London, UK: John Murray; 1868

[47] Gon SM, Price EO. Invertebrate domestication: Behavioral considerations. Bioscience. 1984;**34**:575-579

[48] Gaston KJ. The magnitude of global insect species richness. Conservation Biology. 1991;**5**:283-296

[49] Stork NE, McBroom J, Gely C, Hamilton AJ. New approaches narrow global species estimates for beetles, insects, and terrestrial arthropods. Proceedings of the National Academy of Sciences of the United States of America. 2015;**112**:7519-7523

[50] Aswartha HR. Silkworm (*Bombyx mori*) and its constituents: A fascinating insect in science and research. Journal of Entomology and Zoology Studies. 2017;**5**:1701-1705

[51] Chauhan TPS, Tayal MK. Mulberry sericulture. In: Omkar, editor. Industrial Entomology. Singapore: Springer; 2017. pp. 197-263

[52] Goldsmith MR, Shimada T, Abe H. The genetics and genomics of the silkworm, *Bombyx mori*. Annual Review of Entomology. 2005;**50**:71-100

[53] Willis J, Wilkins A, Goldsmith M. A brief history of Lepidoptera as model systems. In: Goldsmith MR, Wilkins AS, editors. Molecular Model Systems in the Lepidoptera. Cambridge, MA: Cambridge University Press; 1995. pp. 1-20

[54] Meng X, Zhu F, Chen K. Silkworm: A promising model organism in life science. Journal of Insect Science. 2017;**17**

[55] Nwibo DD, Hamamoto H, Matsumoto Y, Kaito C, Sekimizu K. Current use of silkworm larvae (*Bombyx mori*) as an animal model in pharmaco-medical research. Drug Discoveries & Therapeutics. 2015;**9**:133-135

[56] Damodaram KJP, Kempraj V, Aurade RM, Rajasekhar SB, Venkataramanappa RK, Nandagopal B, et al. Centuries of domestication has not impaired oviposition site-selection function in the silkmoth, *Bombyx mori*. Scientific Reports. 2014;**4**:7472

[57] Xiang H, Liu X, Li M, Zhu YY, Wang L, Cui Y, et al. The evolutionary road from wild moth to domestic silkworm. Nature Ecology & Evolution. 2018;**2**:1268-1279

[58] van Huis A. Potential of insects as food and feed in assuring food security. Annual Review of Entomology. 2013;**58**:563-583

[59] Yang S-Y, Han M-J, Kang L-F, Li Z-W, Shen Y-H, Zhang Z. Demographic history and gene flow during silkworm domestication. BMC Evolutionary Biology. 2014;**14**:185

[60] Sun W, Yu H, Shen Y, Banno Y, Xiang Z, Zhang Z. Phylogeny and evolutionary history of the silkworm. Science China. Life Sciences. 2012;**55**:483-496

[61] Xia Q, Guo Y, Zhang Z, Li D, Xuan Z, Li Z, et al. Complete resequencing of 40 genomes reveals domestication events and genes in silkworm (*Bombyx*). Science. 2009;**326**:433-436

[62] Liu Y, Li Y, Li X, Qin L. The origin and dispersal of the domesticated Chinese oak silkworm, *Antheraea pernyi*, in China: A reconstruction based on ancient texts. Journal of Insect Science. 2010;**10**:180

[63] Arunkumar KPP, Metta M, Nagaraju J. Molecular phylogeny of silkmoths reveals the origin of domesticated silkmoth, *Bombyx mori* from Chinese *Bombyx mandarina* and paternal inheritance of *Antheraea proylei* mitochondrial DNA. Molecular Phylogenetics and Evolution. 2006;**40**:419-427

[64] Nakamura T, Banno Y, Fujii H. Genetics of the "wild silkworm translucent" mutant (ows) discovered in the progenies after the cross between the domesticated silkworm, *Bombyx mori*, and the wild mulberry silkworm, *Bombyx mandarina*. International Journal of Wild Silkmoth & Silk. 2001;**6**:7-10

[65] Yukuhiro K, Sezutsu H, Tamura T, Kosegawa E, Iwata K, Ajimura M, et al. Little gene flow between domestic silkmoth *Bombyx mori* and its wild relative *Bombyx mandarina* in Japan, and possible artificial selection on the CAD gene of *B. mori*. Genes & Genetic Systems. 2012;**87**:331-340

[66] Liu X. The Silk Road in World History. Oxford, UK: Oxford University Press; 2010

[67] Fang S-M, Hu B-L, Zhou Q-Z, Yu Q-Y, Zhang Z. Comparative analysis of the silk gland transcriptomes between the domestic and wild silkworms. BMC Genomics. 2015;**16**:60

[68] Yu H-S, Shen Y-H, Yuan G-X, Hu Y-G, Xu H-E, Xiang Z-H, et al. Evidence of selection at melanin synthesis pathway loci during silkworm domestication. Molecular Biology and Evolution. 2011;**28**:1785-1799

[69] Bisch-Knaden S, Daimon T, Shimada T, Hansson BS, Sachse S. Anatomical and functional analysis of domestication effects on the olfactory system of the silkmoth *Bombyx mori*. Proceedings of the Royal Society B: Biological Sciences. 2013;**281**:20132582

[70] Sakudoh T, Nakashima T, Kuroki Y, Fujiyama A, Kohara Y, Honda N, et al. Diversity in copy number and structure of a silkworm morphogenetic gene as a result of domestication. Genetics. 2011;**187**:965-976

[71] Takai H, Asaoka K, Ishizuna F, Kiuchi T, Katsuma S, Shimada T. Morphological and electrophysiological differences in tarsal chemosensilla between the wild silkmoth *Bombyx mandarina* and the domesticated species *Bombyx mori*. Arthropod Structure & Development. 2018;**47**:238-247

[72] Zanatta DB, Bravo JP, Barbosa JF, Munhoz REF, Fernandez MA. Evaluation of economically important traits from sixteen parental strains of the silkworm *Bombyx mori* L (Lepidoptera: Bombycidae). Neotropical Entomology. 2009;**38**:327-331

[73] Defoliart GR. Edible insects as minilivestock. Biodiversity and Conservation. 1995; **4**:306-321

[74] Singh BK, Ahmed SA. Ericulture. In: Omkar, editor. Industrial Entomology. Singapore: Springer; 2017. pp. 345-370

[75] Michener CD. The Bees of the World. Baltimore, MD: The Johns Hopkins University Press; 2000

[76] Engel MS. The taxonomy of recent and fossil honey bees. Journal of Hymenoptera Research. 1999;**8**:165-196

[77] Oldroyd BP, Nanork P. Conservation of Asian honey bees. Apidologie. 2009;**40**:296-312

[78] Hepburn HR, Radloff SE. Biogeography of the dwarf honeybees, *Apis andreniformis* and *Apis florea*. Apidologie. 2011;**42**:293-300

[79] Rattanawannee A, Chanchao C, Wongsiri S. Morphometric and genetic variation of small dwarf honeybees *Apis andreniformis* Smith, 1858 in Thailand. Insect Science. 2007;**14**: 451-460

[80] Ratnayake GM, Weerathunga PN, Dilrukshi MSA, Amara Witharana EWR, Jayasinghe S. Giant honey bee (*Apis dorsata*) sting and acute limb ischemia: A case report and review of the literature. BMC Research Notes. 2018;**11**:327

[81] Garnery L, Vautrin D, Cornuet JM, Solignac M. Phylogenetic relationships in the genus *Apis* inferred from mitochondrial DNA sequence data. Apidologie. 1991;**22**:87-92

[82] Raghunandan KS. Colony status of Asian giant honeybee, *Apis dorsata* Fabricius in Southern Karnataka, India. African Journal of Agricultural Research. 2013;**8**:680-689

[83] Matias DMS, Borgemeister C, von Wehrden H. Ecological changes and local knowledge in a giant honey bee (*Apis dorsata* F.) hunting community in Palawan, Philippines. Ambio. 2018. https://doi.org/10.1007/s13280-018-1038-7

[84] Koeniger N, Kurze C, Phiancharoen M, Koeniger G. "Up" or "down" that makes the difference. How giant honeybees (*Apis dorsata*) see the world. PLoS One. 2017;**12**:e0185325

[85] Paar J, Oldroyd BP, Huettinger E, Kastberger G. Genetic structure of an *Apis dorsata* population: The significance of migration and colony aggregation. The Journal of Heredity. 2004;**95**:119-126

[86] Koeniger N, Koeniger G, Tingek S, Kelitu A. Interspecific rearing and acceptance of queens between *Apis cerana* Fabricius, 1793 and *Apis koschevnikovi* Buttel-Reepen, 1906. Apidologie. 1996;**27**:371-380

[87] Hadisoesilo S, Raffiudin R, Susanti W, Atmowidi T, Hepburn C, Radloff SE, et al. Morphometric analysis and biogeography of *Apis koschevnikovi* Enderlein (1906). Apidologie. 2008;**39**:495-503

[88] Neff JL, Simpson BB. Bees, pollination systems and plant diversity. In: LaSalle J, Gauld ID, editors. Hymenoptera and Biodiversity. Wallingford, UK: CAB International; 1993. pp. 143-167

[89] Yadav S, Kumar Y, Lal Jat B. Honeybee: Diversity, castes and life cycle. In: Omkar, editor. Industrial Entomology. Singapore: Springer; 2017. pp. 5-35

[90] Whitfield CW, Behura SK, Berlocher SH, Clark AG, Johnston JS, Sheppard WS, et al. Thrice out of Africa: Ancient and recent expansions of the honey bee, *Apis mellifera*. Science. 2006;**314**:642-645

[91] Maximino C, Silva RX do C, da Silva S de NS, Rodrigues L do SDS, Barbosa H, de Carvalho TS, et al. Non-mammalian models in behavioral neuroscience: Consequences for biological psychiatry. Frontiers in Behavioral Neuroscience. 2015;**9**:233

[92] Aizen MA, Harder LD. Geographic variation in the growth of domesticated honey bee stocks: Disease or economics? Communicative & Integrative Biology. 2009;**2**:464-466

[93] Crane E. Honey from honeybees and other insects. Ethology Ecology and Evolution. 1991;**3**:100-105

[94] Turner JS. Semiotics of a superorganism. Biosemiotics. 2016;**9**:85-102

[95] Page RE. The evolution of multiple mating behavior by honey bee queens (*Apis mellifera* L.). Genetics. 1980;**96**:263-273

[96] Dadant C. Beekeeping Equipment. The Hive and the Honey Bee. Dadant & Sons: Hamilton, IL; 1975

[97] Laidlaw HH. Artificial insemination of the queen bee (*Apis mellifera* L.): Morphological basis and results. Journal of Morphology 1944;**74**:429-465

[98] Garnery L, Cornuet JM, Solignac M. Evolutionary history of the honey bee *Apis mellifera* inferred from mitochondrial DNA analysis. Molecular Ecology. 1992;**1**:145-154

[99] Arias MC, Sheppard WS. Molecular phylogenetics of honey bee subspecies (*Apis mellifera* L.) inferred from mitochondrial DNA sequence. Molecular Phylogenetics and Evolution. 1996;**5**:557-566

[100] Han F, Wallberg A, Webster MT. From where did the Western honeybee (*Apis mellifera*) originate? Ecology and Evolution. 2012;**2**:1949-1957

[101] Wallberg A, Han F, Wellhagen G, Dahle B, Kawata M, Haddad N, et al. A worldwide survey of genome sequence variation provides insight into the evolutionary history of the honeybee *Apis mellifera*. Nature Genetics. 2014;**46**:1081-1088

[102] Cridland JM, Tsutsui ND, Ramírez SR. The complex demographic history and evolutionary origin of the western honey bee, *Apis mellifera*. Genome Biology and Evolution. 2017;**9**:457-472

[103] Parker R, Melathopoulos AP, White R, Pernal SF, Guarna MM, Foster LJ. Ecological adaptation of diverse honey bee (*Apis mellifera*) populations. PLoS One. 2010;**5**:e11096

[104] Roffet-Salque M, Regert M, Evershed RP, Outram AK, Cramp LJE, Decavallas O, et al. Widespread exploitation of the honeybee by early Neolithic farmers. Nature. 2015;**527**:226-230

[105] Crane E. The World History of Beekeeping and Honey Hunting. London, UK: Routledge; 1999

[106] Harissis HV, Harissis AV. Apiculture in the Prehistoric Aegean. Minoan and Mycenaean Symbols Revisited. Oxford, UK: BAR Publishing; 2009

[107] Ravoet J, De Smet L, Wenseleers T, de Graaf DC. Vertical transmission of honey bee viruses in a Belgian queen breeding program. BMC Veterinary Research. 2015;**11**:61

[108] Witherell PC. A story of success. The starline and midnite hybrid bee breeding programs. American Bee Journal. 1976;**116**:63-64

[109] Niño EL, Cameron JW. Improving the future of honey bee breeding programs by employing recent scientific advances. Current Opinion in Insect Science. 2015;**10**:163-169

[110] Oxley PR, Oldroyd BP. The genetic architecture of honeybee breeding. Advances in Insect Physiology. 2010;**39**:83-118

[111] Adam B. In Search of the Best Strains of Bees. 2nd ed. Hebden Bridge, UK: Peacock Press; 2000

[112] Francoy TM, Wittmann D, Drauschke M, Müller S, Steinhage V, Bezerra-Laure MAF, et al. Identification of Africanized honey bees through wing morphometrics: Two fast and efficient procedures. Apidologie. 2008;**39**:488-494

[113] Harpur BA, Minaei S, Kent CF, Zayed A. Management increases genetic diversity of honey bees via admixture. Molecular Ecology. 2012;**21**:4414-4421

[114] Oldroyd BP. Domestication of honey bees was associated with expansion of genetic diversity. Molecular Ecology. 2012;**21**:4409-4411

[115] Aizen MA, Harder LD. The global stock of domesticated honey bees is growing slower than agricultural demand for pollination. Current Biology. 2009;**19**:915-918

[116] Winfree R, Williams NM, Dushoff J, Kremen C. Native bees provide insurance against ongoing honey bee losses. Ecology Letters. 2007;**10**:1105-1113

[117] Graystock P, Blane EJ, McFrederick QS, Goulson D, Hughes WO. Do managed bees drive parasite spread and emergence in wild bees? International Journal for Parasitology: Parasites and Wildlife. 2015;**5**:64-75

[118] Bailey L. Honey Bee Pathology. London, UK: Academic Press Limited; 1991

[119] De la Rúa P, Jaffé R, Muñoz I, Serrano J, Moritz RFA, Kraus FB. Conserving genetic diversity in the honeybee: Comments on Harpur et al. (2012). Molecular Ecology. 2013;**22**:3208-3210

[120] De la Rúa P, Paxton RJ, Moritz RFA, Roberts S, Allen DJ, Pinto MA, et al. *Apis mellifera*. The IUCN Red List of Threatened Species. 2014. e.T42463639A42463665. Downloaded on 29 October 2018

[121] Zheng H-Q, Jin S-H, Hu F-L, Pirk CWW. Sustainable multiple queen colonies of honey bees, *Apis mellifera ligustica*. Journal of Apicultural Research. 2009;**48**:284-289

[122] Bellard C, Rysman J-F, Leroy B, Claud C, Mace GM. A global picture of biological invasion threat on islands. Nature Ecology & Evolution. 2017;**1**:1862-1869

[123] Tufto J. Domestication and fitness in the wild: A multivariate view. Evolution. 2017; **71**:2262-2270

[124] Bolstad GH, Hindar K, Robertsen G, Jonsson B, Sægrov H, Diserud OH, et al. Gene flow from domesticated escapes alters the life history of wild Atlantic salmon. Nature Ecology & Evolution. 2017;**1**:124

[125] Lecis R, Pierpaoli M, Birò ZS, Szemethy L, Ragni B, Vercillo F, et al. Bayesian analyses of admixture in wild and domestic cats (*Felis silvestris*) using linked microsatellite loci. Molecular Ecology. 2006;**15**:119-131

[126] Fuller ZL, Niño EL, Patch HM, Bedoya-Reina OC, Baumgarten T, Muli E, et al. Genome-wide analysis of signatures of selection in populations of African honey bees (*Apis mellifera*) using new web-based tools. BMC Genomics. 2015;**16**:518

[127] Jaffé R, Dietemann V, Crewe RM, Moritz RFA. Temporal variation in the genetic structure of a drone congregation area: An insight into the population dynamics of wild African honeybees (*Apis mellifera scutellata*). Molecular Ecology. 2009;**18**:1511-1522

[128] Moritz RFA, Kraus FB, Kryger P, Crewe RM. The size of wild honeybee populations (*Apis mellifera*) and its implications for the conservation of honeybees. Journal of Insect Conservation. 2007;**11**:391-397

[129] Thorp RW. Biodiversity of native bees and crop pollination with emphasis on California. In: Gepts P, Bettinger R, Brush S, Damania A, Famula T, McGuire P, et al., editors. Biodiversity in Agriculture: Domestication, Evolution, and Sustainability. New York, USA: Cambridge University Press; 2012. pp. 526-537

[130] Linder HP. Morphology and the evolution of wind pollination. In: Owen SJ, Rudall PJ, editors. Reproductive Biology. Richmond, UK: Royal Botanic Gardens, Kew.; 1998. pp. 123-135

[131] Bawa KS. Plant-pollinator interactions in tropical rain forests. Annual Review of Ecology and Systematics. 1990;**21**:399-422

[132] Garratt MPD, Breeze TD, Jenner N, Polce C, Biesmeijer JC, Potts SG. Avoiding a bad apple: Insect pollination enhances fruit quality and economic value. Agriculture, Ecosystems and Environment. 2014;**184**:34-40

[133] Anderson SH. The relative importance of birds and insects as pollinators of the New Zealand flora. New Zealand Journal of Ecology. 2003;**27**:83-94

[134] Kearns CA. North American dipteran pollinators: Assessing their value and conservation status. Ecology and Society. 2001;**5**:5

[135] VanEngelsdorp D, Meixner MD. A historical review of managed honey bee populations in Europe and the United States and the factors that may affect them. Journal of Invertebrate Pathology. 2010;**103**:S80-S95

[136] Klein AM, Vaissière BE, Cane JH, Steffan-Dewenter I, Cunningham SA, Kremen C, et al. Importance of pollinators in changing landscapes for world crops. Proceedings of Biological sciences. Agroecology. 2007;**274**:303-313

[137] Byrne A, Fitzpatrick Ú. Bee conservation policy at the global, regional and national levels. Apidologie. 2009;**40**:194-210

[138] Slaa EJ, Sánchez Chaves LA, Malagodi-Braga KS, Hofstede FE. Stingless bees in applied pollination: Practice and perspectives. Apidologie. 2006;**37**:293-315

[139] Velthuis HHW, Van Doorn A. A century of advances in bumblebee domestication and the economic and environmental aspects of its commercialization for pollination. Apidologie. 2006;**37**:421-451

[140] Goulson D. Bumblebees: Their Behaviour and Ecology. Oxford: Oxford University Press; 2003

[141] Greenleaf SS, Kremen C. Wild bee species increase tomato production and respond differently to surrounding land use in Northern California. Biological Conservation. 2006;**133**:81-87

[142] Sladen FWL. The Humble-Bee, its Life History and how to Domesticate it, with Descriptions of all the British Species of *Bombus* and *Psithyrus*. London: MacMillian; 1912

[143] Hasselrot TB. Studies on Swedish bumblebees (genus *Bombus* Latr.), their domestication and biology. Opuscula Entomologica Supplement. 1960;**17**:1-192

[144] Velthuis HHW. The historical background of the domestication of the bumble-bee, *Bombus terrestris*, and its introduction in agriculture. In: Kevan P, Imperatriz Fonseca VL, editors. Pollinating Bees—The Conservation Link between Agriculture and Nature. Sao Paulo, Brasil: Ministry of Environment; 2002

[145] Lhomme P, Sramkova A, Kreuter K, Lecocq T, Rasmont P, Ayasse M. A method for year-round rearing of cuckoo bumblebees (Hymenoptera: Apoidea: *Bombus* subgenus *Psithyrus*). Annales de la Société Entomologique de France. 2013;**49**:117-125

[146] Lecocq T, Coppée A, Michez D, Brasero N, Rasplus J-Y, Valterová I, et al. The alien's identity: Consequences of taxonomic status for the international bumblebee trade regulations. Biological Conservation. 2016;**195**:169-176

[147] Lecocq T, Rasmont P, Harpke A, Schweiger O. Improving international trade regulation by considering intraspecific variation for invasion risk assessment of cocmmercially traded species: The *Bombus terrestris* case. Conservation Letters. 2016;**9**:281-289

[148] Cortopassi-Laurino M, Imperatriz-Fonseca VL, Roubik DW, Dollin A, Heard T, Aguilar I, et al. Global meliponiculture: Challenges and opportunities. Apidologie. 2006;**37**:275-292

[149] Crane E. The past and present status of beekeeping with stingless bees. Bee World. 2015;**73**:29-42

[150] Halcroft MT, Spooner-Hart R, Haigh AM, Heard TA, Dollin A. The Australian stingless bee industry: A follow-up survey, one decade on. Journal of Apicultural Research. 2013;**52**:1-7

[151] Serrano A, Sousa M, Hallett J, Simmonds MSJ, Nesbitt M, Lopes JA. Identification of *Dactylopius* cochineal species with high-performance liquid chromatography and multivariate data analysis. The Analyst. 2013;**138**:6081-6090

[152] Wouters J, Verhecken A. The coccid insect dyes: HPLC and computerized diode-array analysis of dyed yarns. Studies in Conservation. 1989;**34**:189-200

[153] Cardon D. Natural Dyes—Sources, Tradition, Technology and Science. London, UK: Archetype Books; 2010

[154] Schulte EK. Improving biological dyes and stains: Quality testing versus standardization. Biotechnic & Histochemistry. 1994;**69**:7-17

[155] Chávez-Moreno CK, Tecante A, Casas A, Claps LE. Distribution and habitat in Mexico of *Dactylopius* Costa (Hemiptera: Dactylopiidae) and their cacti hosts (Cactaceae: Opuntioideae). Neotropical Entomology. 2010;**40**:62-71

[156] Rodríguez LC, Méndez MA, Niemeyer HM. Direction of dispersion of cochineal (*Dactylopius coccus* Costa) within the Americas. Antiquity. 2001;**75**:73-77

[157] Van Dam AR, Portillo Martinez L, Chavez AJ, May BP. Range wide phylogeography of *Dactylopius coccus* (Hemiptera: Dactylopiidae). Annals of the Entomological Society of America. 2015;**108**:299-310

[158] Casas A, Barbera G. Mesoamerican domestication and diffusion of cacti. In: Nobel SP, editor. Cacti: Biology and Uses. Los Angeles, CA: University of California Press; 2002. pp. 143-162

[159] Ramírez-Puebla ST, Rosenblueth M, Chávez-Moreno CK, de Lyra MCCP, Tecante A, Martínez-Romero E. Molecular phylogeny of the genus *Dactylopius* (Hemiptera: Dactylopiidae) and identification of the symbiotic bacteria. Environmental Entomology. 2010;**39**:1178-1183

[160] Sharma K, Jaiswal A, Kk K. Role of lac culture in biodiversity conservation: Issues at stake and conservation strategy. Current Science. 2006;**91**:894-897

[161] Mishra YD, Kumar P. Lac culture. In: Omkar, editor. Industrial Entomology. Singapore: Springer; 2017. pp. 109-155

[162] Ahmad A, Kaushik S, Ramamurthy VV, Lakhanpaul S, Ramani R, Sharma KK, et al. Mouthparts and stylet penetration of the lac insect *Kerria lacca* (Kerr) (Hemiptera: Tachardiidae). Arthropod Structure & Development. 2012;**41**:435-441

[163] Payne CLR, Van Itterbeeck J. Ecosystem services from edible insects in agricultural systems: A review. Insects. 2017;**8**:24

[164] Mishra G, Omkar. Insects as food. In: Omkar, editor. Industrial Entomology. Singapore: Springer; 2017. pp. 413-434

[165] Stamer A. Insect proteins—A new source for animal feed: The use of insect larvae to recycle food waste in high-quality protein for livestock and aquaculture feeds is held back largely owing to regulatory hurdles. EMBO Reports. 2015;**16**:676-680

[166] Dossey AT. Insects and their chemical weaponry: New potential for drug discovery. Natural Product Reports. 2010;**27**:1737-1757

[167] Ratcliffe NA, Mello CB, Garcia ES, Butt TM, Azambuja P. Insect natural products and processes: New treatments for human disease. Insect Biochemistry and Molecular Biology. 2011;**41**:747-769

[168] Jäch MA. Fried water beetles cantonese style. American Entomologist. 2003;**49**:34-37

[169] Halloran A, Roos N, Flore R, Hanboonsong Y. The development of the edible cricket industry in Thailand. Journal of Insects as Food and Feed. 2016;**2**:91-100

[170] Szelei J, Woodring J, Goettel MS, Duke G, Jousset F-X, Liu KY, et al. Susceptibility of North-American and European crickets to *Acheta domesticus densovirus* (AdDNV) and associated epizootics. Journal of Invertebrate Pathology. 2011;**106**:394-399

[171] Kok R, Lomaliza K, Shivhare US. The design and performance of an insect farm/chemical reactor for human food production. Canadian Agricultural Engineering. 1988;**30**:307-317

[172] Van Itterbeeck J, van Huis A. Environmental manipulation for edible insect procurement: A historical perspective. Journal of Ethnobiology and Ethnomedicine. 2012;**8**:3

[173] Payne CLR, Evans JD. Nested houses: Domestication dynamics of human–wasp relations in contemporary rural Japan. Journal of Ethnobiology and Ethnomedicine. 2017;**13**:13

[174] Food and Agriculture Organization of the United Nations. The State of Food Insecurity in the World—The Multiple Dimensions of Food Security. Rome, Italy: FAO; 2013

[175] Thurman JH, Crowder DW, Northfield TD. Biological control agents in the Anthropocene: Current risks and future options. Current Opinion in Insect Science. 2017;**23**: 59-64

[176] Stenberg JA, Heil M, Åhman I, Björkman C. Optimizing crops for biocontrol of psts and disease. Trends in Plant Science. 2015;**20**:698-712

[177] Hoddle MS, Van Driesche RG, Sanderson JP. Biology and use of the whitefly parasitoid *Encarsia formosa*. Annual Review of Entomology. 1998;**43**:645-669

[178] Sithanantham S, Ballal CR, Jalali SK, Bakthavatsalam N. In: Sithanantham S, Ballal CR, Jalali SK, Bakthavatsalam N, editors. Biological Control of Insect Pests Using Egg Parasitoids [Internet]. Springer India: New Delhi; 2013

[179] Kapranas A, Tena A. Encyrtid parasitoids of soft scale insects: Biology, behavior, and their use in biological control. Annual Review of Entomology. 2015;**60**:195-211

[180] Liu T-X, Stansly PA, Gerling D. Whitefly parasitoids: Distribution, life history, bionomics, and utilization. Annual Review of Entomology. 2015;**60**:273-292

[181] Knoll V, Ellenbroek T, Romeis J, Collatz J. Seasonal and regional presence of hymenopteran parasitoids of *Drosophila* in Switzerland and their ability to parasitize the invasive *Drosophila suzukii*. Scientific Reports. 2017;**7**:40697

[182] Smith SM. Biological control with *Trichogramma*: Advances, successes, and potential of their use. Annual Review of Entomology. 1996;**41**:375-406

[183] Weldon CW. Mass-rearing and sterilisation alter mating behaviour of male Queensland fruit fly, *Bactrocera tryoni* (Froggatt) (Diptera: Tephritidae). Australian Journal of Entomology. 2005;**44**:158-163

[184] Kakinohana H, Kuba H, Kohama T, Kinjo K, Taniguchi M, Nakamori H, et al. Eradication of the melon fly, *Bactrocera cucurbitae* coquillett, by mass release of sterile flies in Okinawa Prefecture, Japan. Japan Agricultural Research Quarterly. 1997;**31**:91-100

[185] Rull J, Encarnación N, Birke A. Mass rearing history and irradiation affect mating performance of the male fruit fly, *Anastrepha obliqua*. Journal of Insect Science. 2012;**12**:1-17

[186] Lees RS, Gilles JR, Hendrichs J, Vreysen MJ, Bourtzis K. Back to the future: The sterile insect technique against mosquito disease vectors. Current Opinion in Insect Science. 2015;**10**:156-162

[187] Gilles JRL, Schetelig MF, Scolari F, Marec F, Capurro ML, Franz G, et al. Towards mosquito sterile insect technique programmes: Exploring genetic, molecular, mechanical and behavioural methods of sex separation in mosquitoes. Acta Tropica. 2014;**132**(Suppl):S178-S187

[188] Saeidi Z, Vatandoost H. Aquatic insect from Iran for possible use of biological control of main vector-borne disease of malaria and water indicator of contamination. Journal of Arthropod-Borne Diseases. 2018;**12**:1-15

[189] Sáenz-de-Cabezón FJ, Zalom FG, López-Olguín JF. A review of recent patents on macroorganisms as biological control agents. Recent Patents on Biotechnology. 2010;**4**: 48-64

[190] Bourtzis K, Hendrichs J. Preface: Development and evaluation of improved strains of insect pests for sterile insect technique (SIT) applications. BMC Genetics. 2014;**15**:I1

[191] Bodson L. Motivations for pet-keeping in ancient Greece and Rome: A preliminary survey. In: Podberscek AL, Paul ES, Serpell JA, editors. Companion Animals and Us. Cambridge, MA: Cambridge University Press; 1997. pp. 27-41

[192] Wagler R, Wagler A. Arthropods: Attitude and incorporation in preservice elementary teachers. International Journal of Environmental Science and Technology. 2011;**6**:229-250

[193] Yager D. Comparative aspects of rearing and breeding mantids. In: Wells P, editor. The Praying Mantids: Research Perspectives. Baltimore, MD: Johns Hopkins University Press; 1999. pp. 311-317

[194] Headstrom R. Your Insect Pet. New York, NY: David McKay Co.; 1973

[195] Hu Y, Hu S, Wang W, Wu X, Marshall FB, Chen X, et al. Earliest evidence for commensal processes of cat domestication. Proceedings of the National Academy of Sciences of the United States of America. 2014;**111**:116-120

[196] Tiemann I, Rehkämper G. Evolutionary pets: Offspring numbers reveal speciation process in domesticated chickens. PLoS One. 2012;**7**:e41453

[197] Young MS. The evolution of domestic pets and companion animals. The Veterinary Clinics of North America. Small Animal Practice. 1985;**15**:297-309

[198] Phillips JF, Lockey RF. Exotic pet allergy. Journal of Allergy and Clinical Immunology. 2009;**123**:513-515

[199] Yoder JA, Glenn BD, Benoit JB, Zettler LW. The giant Madagascar hissing-cockroach (*Gromphadorhina portentosa*) as a source of antagonistic moulds: Concerns arising from its use in a public setting. Mycoses. 2007;**51**:95-98

[200] Morgan MS, Arlian LG, Bernstein JA, Yoder JA. Allergenicity of the Madagascar hissing cockroach. Annals of Allergy, Asthma and Immunology. 2007;**98**:258-261

[201] Thomas MC. Invertebrate pets and the Florida department of agriculture and consumer services. Florida Entomologist. 1995;**78**:39-44

[202] Boucher S, Varady-Szabo H. Effects of different diets on the survival, longevity and growth rate of the Annam stick insect, *Medauroidea extradentata* (Phasmatodea: Phasmatidae). Journal of Orthoptera Research. 2005;**14**:115-118

[203] Carey AF, Carlson JR. Insect olfaction from model systems to disease control. Proceedings of the National Academy of Sciences of the United States of America. 2011; **108**:12987-12995

[204] Honeybee Genome Sequencing Consortium. Insights into social insects from the genome of the honeybee *Apis mellifera*. Nature. 2006;**443**:931-949

[205] Shaw AE, Veronesi E, Maurin G, Ftaich N, Guiguen F, Rixon F, et al. *Drosophila melanogaster* as a model organism for bluetongue virus replication and tropism. Journal of Virology. 2012;**86**:9015-9024

[206] Schneider D. Using *Drosophila* as a model insect. Nature Reviews. Genetics. 2000;**1**: 218-226

[207] Sun Y, Yolitz J, Wang C, Spangler E, Zhan M, Zou S. Aging studies in *Drosophila melanogaster*. Methods in Molecular Biology. 2013;**1048**:77-93

[208] Kohler RE. Drosophila: A life in the laboratory. Journal of the History of Biology. 1993;**26**:281-310

[209] Allen G. The introduction of *Drosophila* into the study of heredity and evolution: 1900-1910. Isis. 1975;**66**:322-333

[210] Lewis E. A new standard food medium. Drosophila Information Service. 1960;**34**:117-118

[211] Lakovaara S. Malt as a culture medium for *Drosophila* species. Drosophila Information Service. 1969;**44**:128

[212] Drosophila AM. A Laboratory Handbook. Cold Spring Harbor, NY: Cold Spring Harbor Laboratory Press; 1989

[213] Rose MR. Laboratory evolution of postponed senescence in *Drosophila melanogaster*. Evolution. 1984;**38**:1004-1010

[214] Hoffmann AA, Hallas R, Sinclair C, Mitrovski P. Levels of variation in stress resistance in *Drosophila* among strains, local populations, and geographic regions: Patterns for desiccation, starvation, cold resistance, and associated traits. Evolution. 2007;**55**:1621-1630

[215] Cortese MD, Norry FM, Piccinali R, Hasson E. Direct and correlated responses to artificial selection on developmental time and wing length in *Drosophila buzzatii*. Evolution. 2002;**56**:2541-2547

[216] Harshman LG, Moore KM, Sty MA, Magwire MM. Stress resistance and longevity in selected lines of *Drosophila melanogaster*. Neurobiology of Aging. 1999;**20**:521-529

[217] Scott D, Shields A, Straker M, Dalrymple H, Dhillon PK, Harbinder S. Variation in the male pheromones and mating success of wild caught *Drosophila melanogaster*. PLoS One. 2011;**6**:e23645

[218] Stanić S, Pavković-Lucic S. Mating success of wild type and sepia mutants *Drosophila melanogaster* in different choice. Rivista di Biologia. **98**:513-524

[219] Simões P, Rose MR, Duarte A, Gonçalves R, Matos M. Evolutionary domestication in *Drosophila subobscura*. Journal of Evolutionary Biology. 2007;**20**:758-766

[220] Stanley CE, Kulathinal RJ. Genomic signatures of domestication on neurogenetic genes in *Drosophila melanogaster*. BMC Evolutionary Biology. 2016;**16**:6

[221] Sgrò CM, Partridge L. Evolutionary responses of the life history of wild-caught *Drosophila melanogaster* to two standard methods of laboratory culture. The American Naturalist. 2000;**156**:341-353

[222] Krebs RA, Roberts SP, Bettencourt BR, Feder ME. Changes in thermotolerance and Hsp70 expression with domestication in *Drosophila melanogaster*. Journal of Evolutionary Biology. 2001;**14**:75-82

[223] Hoffmann AA, Hallas R, Sinclair C, Partridge L. Rapid loss of stress resistance in *Drosophila melanogaster* under adaptation to laboratory culture. Evolution. 2001;**55**: 436-438

[224] Matos M, Simões P, Duarte A, Rego C, Avelar T, Rose MR. Convergence to a novel environment: Comparative method versus experimental evolution. Evolution. 2004; **58**:1503-1510

[225] Versace E, Nolte V, Pandey RV, Tobler R, Schlötterer C. Experimental evolution reveals habitat-specific fitness dynamics among Wolbachia clades in *Drosophila melanogaster*. Molecular Ecology. 2014;**23**:802-814

[226] Tobler R, Franssen SU, Kofler R, Orozco-Terwengel P, Nolte V, Hermisson J, et al. Massive habitat-specific genomic response in *D. melanogaster* populations during experimental evolution in hot and cold environments. Molecular Biology and Evolution. 2014;**31**:364-375

[227] Santos M, Fragata I, Santos J, Simões P, Marques A, Lima M, et al. Playing Darwin. Part B. 20 years of domestication in *Drosophila subobscura*. Theory in Biosciences. 2010;**129**:97-102

[228] Pool JE, Corbett-Detig RB, Sugino RP, Stevens KA, Cardeno CM, Crepeau MW, et al. Population genomics of sub-Saharan *Drosophila melanogaster*: African diversity and non-African admixture. PLoS Genetics. 2012;**8**:e1003080

[229] Latter BD, Mulley JC. Genetic adaptation to captivity and inbreeding depression in small laboratory populations of *Drosophila melanogaster*. Genetics. 1995;**139**:255-266

[230] Sgrò CM, Geddes G, Fowler K, Partridge L. Selection on age at reproduction in *Drosophila melanogaster*: Female mating frequency as a correlated response. Evolution. 2000;**54**:2152-2155

[231] De Queiroz K. Ernst Mayr and the modern concept of species. Proceedings of the National Academy of Sciences of the United States of America. 2005;**102**(Suppl):6600-6607

[232] De Queiroz K. Species concepts and species delimitation. Systematic Biology. 2007; **56**:879-886

[233] Lecocq T, Dellicour S, Michez D, Dehon M, Dewulf A, De Meulemeester T, et al. Methods for species delimitation in bumblebees (Hymenoptera, Apidae, *Bombus*): Towards an integrative approach. Zoologica Scripta. 2015;**44**:281-297

[234] Omkar. Industrial Entomology. Singapore: Springer Singapore; 2017. 465 pp

[235] Raghavan R, Dahanukar N, Tlusty MF, Rhyne AL, Krishna Kumar K, Molur S, et al. Uncovering an obscure trade: Threatened freshwater fishes and the aquarium pet markets. Biological Conservation. 2013;**164**:158-169

[236] Germain N, Hartmann HJ, Fernández-Rivera Melo FJ, Reyes-Bonilla H. Ornamental reef fish fisheries: New indicators of sustainability and human development at a coastal community level. Ocean and Coastal Management. 2015;**104**:136-149

[237] Dhar B, Ghosh SK. Genetic assessment of ornamental fish species from North East India. Gene. 2015;**555**:382-392

[238] DeHaan LR, Van Tassel DL, Anderson JA, Asselin SR, Barnes R, Baute GJ, et al. A pipeline strategy for grain crop domestication. Crop Science. 2016;**56**:917

[239] Williams SE, Hoffman EA. Minimizing genetic adaptation in captive breeding programs: A review. Biological Conservation. 2009;**142**:2388-2400

Chapter 4

Fish Domestication: An Overview

Teletchea Fabrice

Additional information is available at the end of the chapter

http://dx.doi.org/10.5772/intechopen.79628

Abstract

The first trials of fish farming date back about 3500 years. Yet, this is only in the 1980s that the exponential development of aquaculture occurred, driven by the strong demand for aquatic products, whereas global capture fisheries started decreasing. The strong increase of aquaculture production has relied chiefly on the domestication of an increasing number of fish species. Nevertheless, only a limited number has reached a high level of domestication. The aim of this chapter is to describe the domestication history of five of the main farmed fish species globally in which all have reached Level 5 and one example that has reached Level 2 only. These six species were chosen because enough scientific literature was available to describe their domestication history that allows illustrating both ancient and recent fish domestication.

Keywords: domestication level, wild, domesticated, fish species, aquaculture, diversification

1. Introduction

The first trials of fish farming date back about 3500 years [1, 2]. In Egypt, the earliest evidence of some kind of control of reproduction of the Nile tilapia (*Oreochromis niloticus*) in irrigation ponds is suggested about 1500 years BC on paintings and bas-reliefs found in Theban tombs [3, 4]. Nearly at the same period, the farming of the common carp (*Cyprinus carpio*) started in China, in link with irrigation of agriculture [1–3]. This is also in China that the first treatise ever written on aquaculture "Fish breeding" was published in 475 years BC by Fan Li [1, 5–7]. Few centuries later, the farming of three other carps started in China: the silver carp (*Hypophthalmichthys molitrix*), the bighead carp (*Hypophthalmichthys nobilis*) and the grass carp (*Ctenopharyngodon idella*). Also in Asia, the aquaculture in brackish water has arguably started in the fifteenth century with the farming of milkfish (*Chanos chanos*) in Java [6]. In Europe, the farming of the common carp in freshwater ponds was already well developed during the Middle Ages [5, 8, 9]. The Italian "Vallicoltura" (farming in coastal lagoons and large open waterbodies of seabass, sea bream, eel, mullets, etc.) dates back to the fifteenth century, and the French trout culture started developing around the second half of the nineteenth century [9]. The rearing of oysters is also very ancient (dating back to 2500 years BC), either in Europe (particularly in Greece) or in Japan [9]. Nevertheless, other regions have a much more recent history of aquaculture. In North America, aquaculture started about 100 years ago [9]. In Africa, the aquaculture production dates back to the 1940s. In Australia, New-Zealand, and diverse Pacific Island states, the development of aquaculture is even more recent [9]. In conclusion, the rearing of aquatic animals is very old [10], particularly in Asia [11], even though the global increase is much more recent and has occurred mainly in the past three decades [4, 9, 11].

In 1950, global aquaculture production was almost negligible (about half a million tons), particularly in comparison to capture fisheries [9]. Up until the end of the 1970s, the production only slightly increased and represented at that time about 6% of the global production of aquatic products [2]. Yet, at the beginning of the 1980s, the improvements of rearing conditions, animal nutrition, larval rearing, and more recently genetics allowed strongly improving the production of an increasing number of species [9, 12, 13]. The exponential development of aquaculture during this period of time (**Figure 1**) was driven by the strong demand for aquatic products, whereas global capture fisheries started decreasing [14]. Between 1980 and 2010, global aquaculture production was multiplied by 12, with a mean annual increase of 8.8%, sometimes reaching more than 12% during certain years [9, 15]. During the 2000s, the aquaculture production has continued increasing [10, 16–18], but at a slower pace (6.8%) than in the 1980s (10.8%) and 1990s (9.5%) [19]. In 2014, aquaculture provided half of all aquatic products for human consumption [20], although this percentage was about 9% in 1980 [19]. More than 500 species and/or species groups had been farmed by 2014 (i.e., a sevenfold increase compared to 1950 [21]), including 362 finfishes, 140 molluscs, 62 crustaceans, and about 15 other aquatic animals [20]. The global aquaculture production (**Figure 1**) is dominated in 2016 by the farming of freshwater fish species (58%, i.e., 46.4 million tons) followed by the production of molluscs (21.4%), crustaceans (9.8%), diadromous fishes (6.2%), marine fishes (3.4%), and various aquatic animals (1.2%).

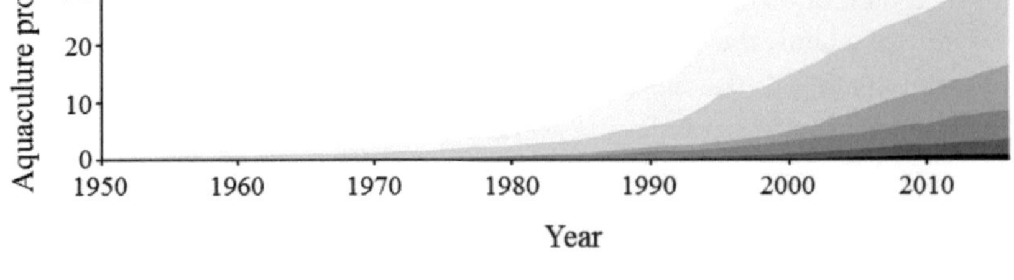

Figure 1. Global aquaculture production (excluding plants) from 1950 to 2016 (data from the FAO database).

Level	Definition	Top three families (number of species)
5	Selective breeding program is used focusing on specific goals	Cyprinidae (10), Salmonidae (8), Acipenseridae (5)
4	Entire life cycle closed in captivity without wild inputs	Cichlidae (6), Sparidae (5), Cyprinidae (4)
3	Entire life cycle closed in captivity with wild inputs	Sparidae (8), Cyprinidae (4), four families (3)
2	Part of the live cycle closed in captivity: several bottlenecks	Cyprinidae (9), Serranidae (5), Carangidae (4)
1	First trials of acclimatization to the captive environment	Cyprinidae (8), Siganidae (3), Sciaenidae (3)

Data from Teletchea and Fontaine [12].

Table 1. Domestication levels with their description and the list of the three leading families by levels in terms of number of species; the number of species by families by level is indicated in parentheses.

The aim of this chapter is to describe the domestication history of five of the main farmed fish species globally in which all have reached Level 5 and one example that has reached Level 2 only (**Table 1**). These six species were chosen because enough scientific literature was available to describe their domestication history that allows illustrating both ancient and recent fish domestication.

2. Domestication of freshwater fish species

2.1. The common carp (*Cyprinus carpio*)

The farming of the common carp, which started in China, is one of the most ancient with the tilapia farming, dating back at least 2500 years [5, 22]. During its long history of domestication, the common carp was profoundly modified by man, deliberately or not, resulting in several breeds with different body shapes, scales, colors, and performances [5, 23–25]. One of the most obvious changes is scales, whose shape and number strongly vary between the four main types of carps: scaled carp, line carp, mirror carp, and leather or naked carps [23–25]. Besides, the body has evolved from an elongated and torpedo shaped for wild individuals to become more laterally compressed and deep bodied [23]. The coloration is now also highly variable, including vivid pigmentations of white, red, black, yellow, blue, gold, and silver [5]. Most modifications of body shape, scalation, and color may have not appeared before the sixteenth century, even though it is possible that some might have existed before [5]. During the nineteenth century, a new trend developed, mainly in Japan [23], with the selection of colored carps [5, 22]. These new forms called "nishikigoi" or simply "Koï" were reproduced, selected, and auctioned for use as pets in small garden pools and many public waterbodies [5, 23]. This intensive selection and the commercial production of these various colored carps have only truly started in the 1950s [5]. Today, the Japanese recognize at least 13 basic color patterns, each with numerous varieties [5]. They would be today more than a hundred varieties of Koi [22], among which some specimens are the most expensive ornamental fish globally:

individual fish can reach a million dollars [23]. However, despite Koi (for the ornamental market), the bulk of global aquaculture production of the common carp for human consumption is based on unselected individuals [24–27]. Indeed, even though carps have undergone selective breeding, at least empirically, for centuries, a more rationale practice has nevertheless been used only at the beginning of the twentieth century [24–27].

During its long history of domestication, the common carp, whose natural range spreads from the Danube River (in Europe) to the Amur River (in China), was introduced in more than 120 countries; its current distribution is now almost global [5, 28–29]. In Europe, the common carp would be native from central regions (particularly the Danube delta), where its farming was started by the Romans about 2000 years ago [5]. During the Middle Ages, its farming spread to the entire European continent [5, 22, 25], particularly due to the development of fish farming in monasteries [5, 30]. In most countries where it was introduced, the common carp was naturalized, establishing self-sustaining populations, and qualified as feral, that is, farmed individuals returning to the wild [5, 23]. These feral populations are today found in hydrosystems of more than 90 countries globally [29], in Asia, Africa, North, Central and South America, Australia, and Oceania [31]. In Australia, for instance, the common carp is the largest freshwater fish introduced and now contributes more than 90% of fish biomass in many areas of south-east of Australia [31]. In Europe, feral populations, some of which could be centuries old, dominate most of drainage systems [22].

Today, the common carp is the third most important species in aquaculture production by volume with more than 4.1 million tons in 2014 (**Figure 2**). Its production was multiplied by more than 40 in the past decades (**Figure 2**). In 2000, it was produced in more than 80 countries, among which about 15 displayed a production higher than 10,000 tons [15, 22]. The rearing systems of the common carp are highly diversified: extensive monoculture in earthen ponds, with or without fertilization and/or additional feeding; intensive monoculture in small ponds, cages, or raceways; and integrated mono- or polyculture with other agricultural activities [32].

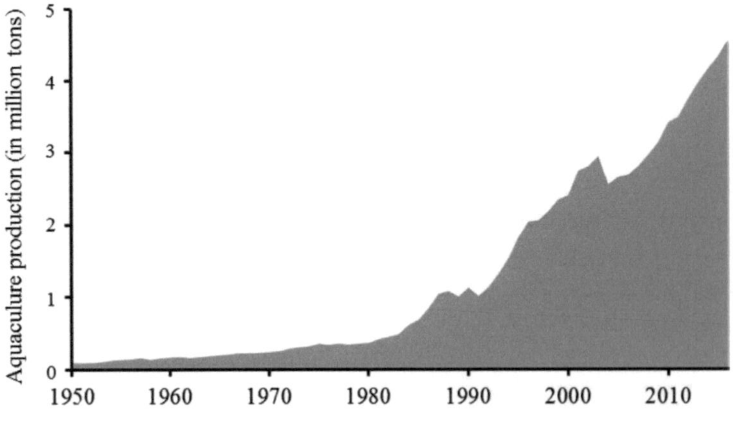

Figure 2. Global aquaculture production from 1950 to 2016 of the common carp *Cyprinus carpio* (data from the FAO database).

In Europe, the common carp is traditionally farmed in large ponds of one to several hundred hectares [32–34]. The degree of farming intensity ranges from highly extensive (low stocking densities with no additional feeding or fertilization) to relatively highly intensive (high density stocking, control of water quality, mechanical aeration, provision of complete feeds, etc.) [32]. In extensive systems, the common carp is the main species reared, sometimes associated with other cyprinids—the tench (*Tinca tinca*), the roach (*Rutilus rutilus*), the rudd (*Scardinius erythrophthalmus*)—and various piscivorous fish—the northern pike (*Esox lucius*), the Eurasian perch (*Perca fluviatilis*), the pikeperch (*Sander lucioperca*), or the Wels fish (*Silurus glanis*) [32, 33]. The main European producing countries are Russia, Ukraine, Czech Republic, Poland, Hungary, and Germany [22, 34]. In Japan and Southeast Asia, the traditional farming of the common carp is practiced in monoculture associated with agriculture (rice and cereals) or farming (duck) [32]. In both China and India, the common carp is traditionally reared in polyculture ponds in association with five or six species [32]: the silver carp (*Hypophthalmichthys molitrix*), the bighead carp (*Hypophthalmichthys nobilis*), the grass carp (*Ctenopharyngodon idella*), the mud carp (*Cirrhinus molitorella*), and the black carp (*Mylopharyngodon piceus*) [32]. This association allows obtaining much higher yields for the diverse farmed species [32].

In conclusion, the common carp is one of the first species to have been domesticated and has become the third leading farmed aquatic species and one of the most introduced worldwide [25–29]. Its biological characteristics (robustness, fast growth, easy propagation, omnivorous feeding regime, ability to readily accept pellets, resistance to disease, and tolerance to a wide range of climatic conditions [35]) make the common carp one of the most popular species [22] that succeeds to adapt to different geographic areas and rearing systems [32].

2.2. The Nile tilapia (*Oreochromis niloticus*)

The farming of the Nile tilapia is one of the most ancient in the world and dates back to the Egyptian antiquity, more than 3500 years ago [3]. Representations of farming are still visible today on the bas-reliefs of Theban tombs [3]. During the following centuries, the rearing systems and farmed tilapias had been slightly modified; thus, the global production remained very low, reaching only about 1500 tons in 1950 (**Figure 3**). However, during the second half of the last century, the production started increasing exponentially, notably from the end of the 1980s (**Figure 3**). The explosion of production is chiefly due to the control of the reproduction in captivity, thanks to the onset of monosex farming, including males only [15]. The farming of monosex male populations is today the rule (males displaying a much higher growth rate than females [36]), which is obtained by sex reversal induced by masculinizing hormones [36], hybridization, or both combined [15]. The production of sex reversal by hormonal treatment of fingerlings is now prohibited in fish destined to human consumption in Europe but is still authorized and largely used in the rest of the world [36]. The second method is the breeding with other species, such as the Mozambique tilapia (*Oreochromis mossambicus*) and the blue tilapia (*Oreochromis aureus*) [6]. In parallel to the sex reversal method, a selective breeding program was initiated at the end of the 1980s, in the Philippines, by an international research institute, among which Filipino and Norwegian researchers [36–38]. This program focused on growth and was called "Genetically Improved Farmed Tilapia" or GIFT [36–38]. In order to ensure a broad genetic variability, broodstock

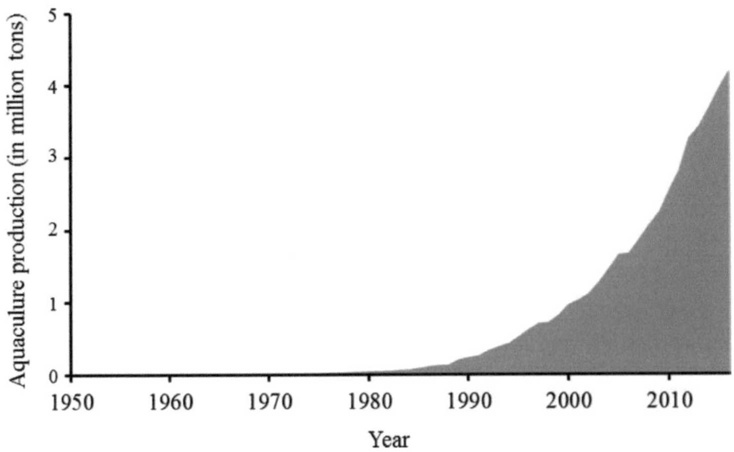

Figure 3. Global aquaculture production from 1950 to 2016 of the Nile tilapia *Oreochromis niloticus* (data from the FAO database).

were sampled from four strains of tilapias reared in the Philippines (Israel, Singapore, Taiwan, and Thailand stocks) as well as four wild populations imported from Africa (Egypt, Ghana, Kenya, and Senegal) [37]. Through five generations of selection, the cumulative gain for growth rate was 86%, corresponding to an average of 17% per generation [37]. At least 20 family-based breeding programs are now in operation globally, more than any other aquatic species [37]. The base populations in 10 of these 20 breeding programs come from the GIFT project, essentially breeders from the fifth generation of selection [37]. The GIFT and GIFT-derived strains are used in numerous farms throughout the world [39, 40]. They represented 80% of the total production of fingerlings in China, 75% in Thailand, and 40% in the Philippines [39]. Today, several species of tilapia are reared in the world, but the Nile tilapia remains by far the dominant species, with more than 80% of the total production, including those of the hybrids [36]. The two other species significantly produced are the Mozambique tilapia and the blue tilapia [36].

In parallel to its domestication, the Nile tilapia has progressively been introduced in numerous countries around the world, mainly during the years 1960–1980 [36]. It is now farmed in more than 100 countries, and its total production has reached more than 2.5 million tons in 2010 (**Figure 3**); this species ranked at the fifth place among the most produced aquatic species in the world in 2014 [38]. This represents a 60-fold increase of the production since 1980 [2, 36, 37]. According to some experts, tilapia has the potential to become the most important aquaculture species in the world in the coming years [37]. Besides, even though the tilapia is native from Africa, more than 70% of global production comes from Asia with more than 1 million tons from China only [9]. In Africa, the production is much lower (about 24%) and comes exclusively from Egypt with more than 550,000 in 2010 [9]. The rearing systems for tilapia are diverse: extensive or semi-intensive monoculture or polyculture in ponds, intensive monoculture in floating cages, in pens, in tanks, or in recirculated systems [9]. The extensive farming of tilapia in ponds is widespread in tropical zones and mostly destined to local consumption [36]. The intensive or hyper-intensive farming (high densities, selectively improved strains, use of pellets) in tanks or floating cages in large lakes or reservoirs are practiced in several countries, such as China or Indonesia [36]. The farming in recirculated system has developed in temperate countries in order to produce tilapias all year long in controlled conditions, even though the production still remains low [9].

The tilapia, described as the aquatic chicken for its speedy and efficient growth, is often considered as a nearly perfect species for aquaculture [2]. Its biological characteristics are indeed particularly adapted to diverse rearing systems: high rusticity, spontaneous reproduction in captivity (multiple spawning), low-trophic level (phytoplankton and detritus), and extreme feeding plasticity [2, 29, 36]. Besides, the production costs are low, less than one dollar per kilo in subtropical and tropical countries that possess adequate temperature for its growth [36].

2.3. The rainbow trout (*Oncorhynchus mykiss*)

The farming of the rainbow trout started in the 1870s in California [41, 42]. Very rapidly, even though this species reproduces only once a year, the selective breeding in hatcheries and the control of the reproductive cycle using modifications of daylength duration allowed producing eggs virtually year round [41, 42]. In the same time, growth rate, disease resistance, and fecundity were all improved from the years 1930 to 1940, thanks to specific breeding programs [41, 42]. More recently, sterile triploid females were produced from thermal or pressure shock on eggs, leading to the farming of monosex females (which mature later than males and have much better flesh quality), which also ensure that if some fish escape from rearing systems into the wild, they will not be able to reproduce in natural conditions [42]. The rearing of triploid females is widely used in the world, notably in France [43].

In the following years after the onset of domestication, the rainbow trout was exported, in form of eggs, in other American States (Virginia, Michigan, notably), then to Japan, the United Kingdom, and Denmark [15, 41, 42, 44, 45]. During the twentieth century, its introduction was continued in numerous countries [15, 41, 42, 44–46] either for human consumption or for restocking of rivers and lakes for recreational fisheries (particularly in the USA, Europe, and Japan). Consequently, the rainbow trout is one of the most introduced fish species globally, being present in more than 90 countries [47]. The production of rainbow trout has strongly increased, particularly during the 1960s and 1970s [42], partly thanks to the development of new extruded pellets [42], from 4400 tons in 1950 to 145,124 tons in 1980, and continued increasing to reach 728,844 tons in 2010 (**Figure 4**). Today, the rainbow trout is produced in more than 70 countries globally [46, 47], among which 15 have a production higher than 15,000 tons [9]. In 2010, Europe was the leading producer of the rainbow trout with more than one-third of the production, followed by South America (mainly Chile), Asia (mainly Iran and Turkey), and Nord America (the United States) [15, 43, 46]. The production of the rainbow trout is consequently much more important in areas where is had been introduced than in its native range in North America (from Alaska to Mexico) [15, 43]. The rainbow is chiefly produced in intensive monoculture systems: individual ponds, concrete raceways, or other types of flow-through tanks (open systems) [18, 34, 42, 46], recirculated water systems, or floating cages either in freshwater or in marine waters [43, 46].

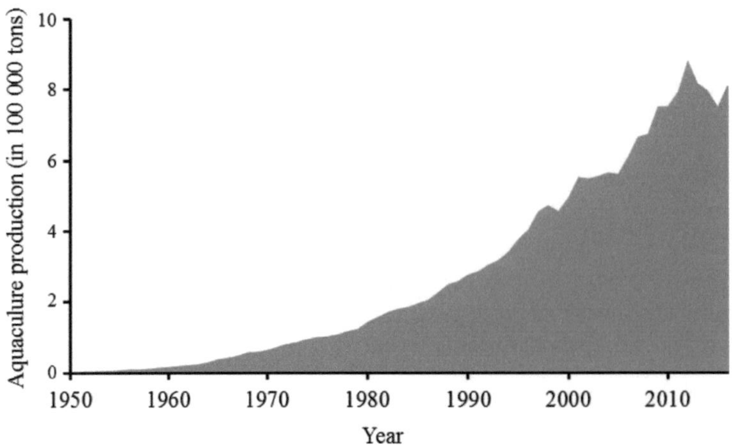

Figure 4. Global aquaculture production from 1950 to 2016 of rainbow trout *Oncorhynchus mykiss* (data from the FAO database).

The rainbow trout is one of the oldest fish in culture [41]. In Europe, up to 14 generations of selection have been performed in family selection and up to 20 in mass selection [44]. This is also the species for which the geographical area was the most increased following the numerous introductions in the past century. In 2014, the rainbow trout was the twelfth most produced aquatic species globally. Its biological characteristics (both sex mature in captivity, spawning is easy to obtain, eggs are relatively robust, fry are sufficiently developed at hatching to directly accept pellets, relatively large tolerance to both temperature and salinity) partly explain the success of its rearing throughout the world [41, 44, 47].

2.4. The striped catfish (*Pangasianodon hypophthalmus*)

The farming of the striped catfish, as well as a dozen of other fish species [48], is very ancient in the Mekong River, particularly in Vietnam, and, to a lesser extent, in Cambodia and Thailand [49]. Traditionally, the farming of the striped catfish relied on the capture of fry and juveniles in the wild, mainly in Cambodia [49], which was then farmed up to a sufficient size to be consumed by local populations [48, 50, 51]. Yet, the farming of striped catfish in Vietnam dramatically changed in the past decades [50], partly due to the ban, by the governments of Cambodia in 1994 and Vietnam in 2000, of the collection of wild fry and juveniles of all species in the natural environment [50], and partly thanks to the development, within the framework of a Franco-Vietnamese research program, of a reliable artificial propagation of two catfish species [49, 50, 51]. The control of reproduction in captivity was first developed for *Pangasius bocourti* [49] and then applied on the striped catfish [50], which allows securing the aquaculture production [49, 52]. In few years, several hundred hatcheries opened in the Mekong Delta allowing to provide all farms [49, 51], which rapidly became the leading farmed species in Vietnam [49]. The control of the reproductive cycle in captivity has changed the aquaculture of this species from being seen as an activity that overexploited natural resources to one that reduces pressure on wild populations [50]. Nevertheless, farming in both cages and pens in Cambodia, Laos, and, to a much lesser extent, Thailand still relies on the capture of juveniles in the wild [9].

http://dx.doi.org/10.5772/intechopen.79628

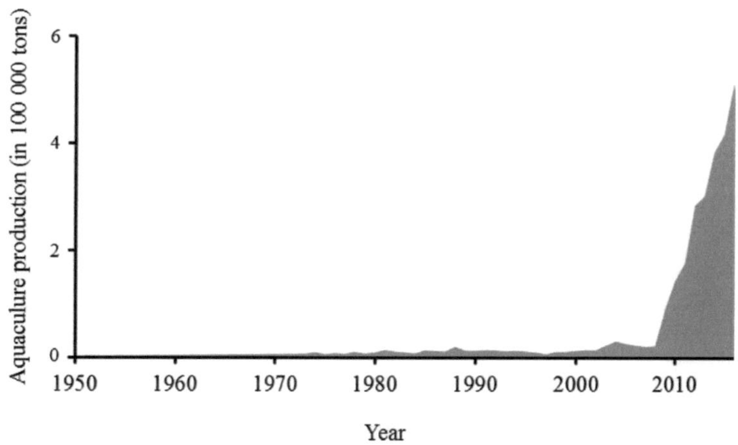

Figure 5. Global aquaculture production from 1950 to 2016 of the striped catfish *Pangasianodon hypophthalmus* (data from the FAO database).

The production of the striped catfish (**Figure 5**) was very low in the 1950s (few thousand tons) and strongly increased from the middle of 1990s [49], to reach more than 1.5 million tons in 2010, including its hybrids [48, 51]. The major part of the production is realized in its native area, i.e., the Mekong River [48, 53]. The Vietnam strongly dominates the production with more than 75% of the global volume, followed by Cambodia, Thailand, and Laos [49, 51]. This species was introduced in neighboring countries, notably in Indonesia, Bangladesh, and Malaysia that produced all together about 20% of the global volume [19, 48, 50]. Only in Vietnam, this species represents more than one third of the aquaculture production [51, 52]. More than 90% of the Vietnamese production of striped catfish is exported to a hundred of countries in the world [49–51, 53, 54]. In parallel to increase of the production, the rearing systems have also evolved [52]. Traditionally, the farming of striped catfish was realized in bamboo floating cages in rivers or reservoirs within which food was mainly constituted of domestic wastes or farming co-products, and in ponds, most often fertilized with organic effluents [49]. Today, most of the production is realized in ponds [52] using commercially made feed [49, 52]. The production could reach hundreds of tons (up to 1000 tons) by ha and year [49].

The striped catfish is without any doubt the aquatic species for which the production has increased the most during the past decade [52, 55]. The domestication of this species is still in its infancy, and genetic improvement programs have only recently started [50]. Because of its low-production cost, mainly due to its feeding regime, its rapid growth, along with high-rearing densities, and large volume, frozen striped catfish fillets have become extremely competitive on the international market [19]. In the early growth phases of the sector, it has to overcome trade embargoes, particularly link to the use of antibiotics; yet, recent studies have demonstrated that there is no food safety concern from either environmental or applied contaminant compounds in this species [49, 52, 56]. Currently, restrictions on the export of the striped catfish from Vietnam do not exist in most importing countries [52]. Catfish fillets are sold in more than 100 countries, among which the United States and more recently Europe (particularly in Eastern countries and Spain) and Asia (mostly Japan, Hong Kong, and Singapore) [19].

3. Domestication of diadromous and marine fish species

3.1. The Atlantic salmon (*Salmo salar*)

The farming of the Atlantic salmon started in the early nineteenth century in the United Kingdom in order to rebuild river populations for angling [9]. Nevertheless, this is only at the end of the 1960s that the farming in sea cages was used for the first time in Norway [57]. From the year 1970, the first ever family-based breeding program of the Atlantic salmon was initiated in Norway in a brand new research institute entitled AKVAFORSK [57, 58]. Even though the production of the Atlantic salmon was only 100 tons in Norway at this time, the equivalent of about 3.6 million US$ was covered for two-thirds by the Norwegian government and the rest by industry (not working in the aquaculture field) and nongovernmental organizations [57]. In parallel, research programs were started to develop new dry pellet feed adapted to the Atlantic salmon, which were available from 1982 [57]. After about 40 years of farming, the time to produce a standard market-sized 4 kg fish has been halved, and while 3 kg of dry matter (in moist feed) was necessary to produce 1 kg of salmon in the beginning, this has also been reduced to 1.15 kg (dry pellets) [37]. Other traits were progressively added to the selection index such as age at sexual maturity, disease resistance, stress resistance, and quality of the flesh [10, 37, 57, 58]. From the early stages of the selective breeding programs, eggs and juveniles were sold to the industry resulting in that close to 100% of the production of the Atlantic salmon in Norway and in the rest of the world are now based on improved stocks [10, 57, 58]. Farmed salmon is regarded as one of the most domesticated fish species farmed for food, and one Norwegian strain has been exposed to ≥12 generations of domestication [59, 60].

In parallel to its domestication, the production was multiplied by 5000 (**Figure 6**) from 294 tons in 1970 to more than 1.4 million tons in 2010 [19, 61]. A dozen of countries currently

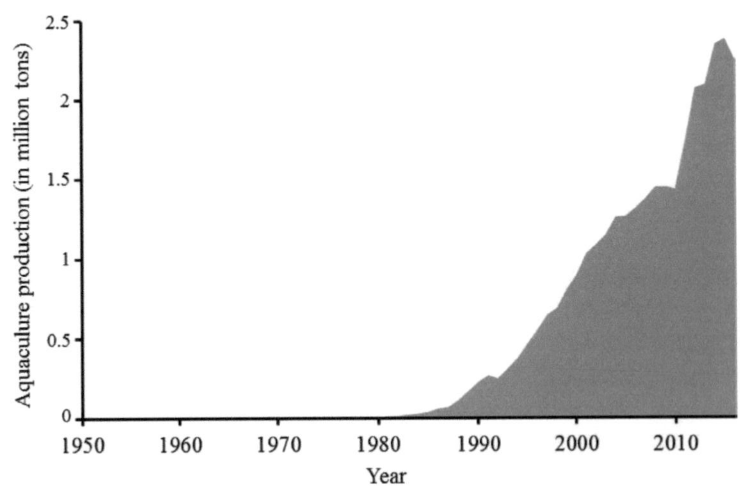

Figure 6. Global aquaculture production from 1950 to 2016 of the Atlantic salmon *Salmo salar* (data from the FAO database).

produce the Atlantic salmon, among which the four leading countries are Norway, Chile, the United Kingdom, and Canada [19, 47, 62]. In Norway and Chile (where this species was introduced), the farming of the Atlantic salmon has an enormous economic importance [63, 64]. In Norway, the farming of Atlantic salmon is the third largest industry after petroleum and mining [47]. In Chile, the exports of the Atlantic salmon represent about two-thirds of the total Chilean fisheries exports and became the third most important export commodity after copper and wood products [47]. Most part of the production (on-growing) is now performed in sea cages [19, 61], with hatcheries working with either flow-through or recirculating aquaculture systems [65].

The Atlantic salmon has evolved in few decades from a luxury item, which was consumed only at specific periods in the year (particularly at Christmas), to convenience products [9]. He has become in 2014 the ninth most produced aquatic products globally, with more than 2.3 million tons. The success of its farming in Norway (and in Chile) is mainly due to the presence of numerous suitable sites for its production, a dynamic research and industry as well as government support [9]. Besides, the Atlantic salmon displays several features (high growth in cages, flesh strongly appreciated by consumers, etc.) that contribute to its farming success [9]. In parallel, the capture by fisheries has strongly decreased from 118,000 tons in 1980 to 73,000 tons in 1990 to less than 40,000 ton in 2000 [47]. Today, almost 100% of the Atlantic salmon is coming from farming [47].

3.2. The European seabass (*Dicentrarchus labrax*)

The traditional farming of the European seabass in the Mediterranean Sea consisted of collecting juveniles in the wild and releasing them in semi-artificial coastal lagoons, such as "Vallicoltura" in Italy [66, 67]. Within these lagoons, seabass was reared extensively [68]. Nevertheless, from the year 1960, in the face of strong competition for wild juveniles between the on-growers and the decrease of natural resources [66], the first rearing trials were initiated in France and Italy [69]. In the middle of the 1980s, the development of reliable methods of reproduction techniques and husbandry methods allowed higher survival [68], and their diffusion rapidly led to the development of a true industry in several countries along the Mediterranean Sea [66, 68]. In the 1990s, the first breeding programs were initiated, first in France and Israel, then in Greece, Spain, and Italy, which allow obtaining the eighth generation of selection in the oldest program [67]. However, a large proportion of the broodstock used today is still coming from wild breeders or first-generation individuals [66, 69, 70].

The aquaculture production of seabass was almost inexistent in 1950 (**Figure 7**). With the control of the life cycle in captivity, the aquaculture production increased exponentially from the middle of 1980 to reach 134,711 tons in 2010 (**Figure 7**). In the same period of time, the capture by fisheries increased from 4460 tons in 1980 to 10,853 tons in 2010 [9]. Consequently, 9 of 10 seabasses consumed in the world are now farmed. The main producers are all located around the Mediterranean Sea: Turkey, Greece, Egypt, Spain, and Italy [9]. Most of the production (on-growing) is realized in sea cages, followed by tanks and lagoon [68].

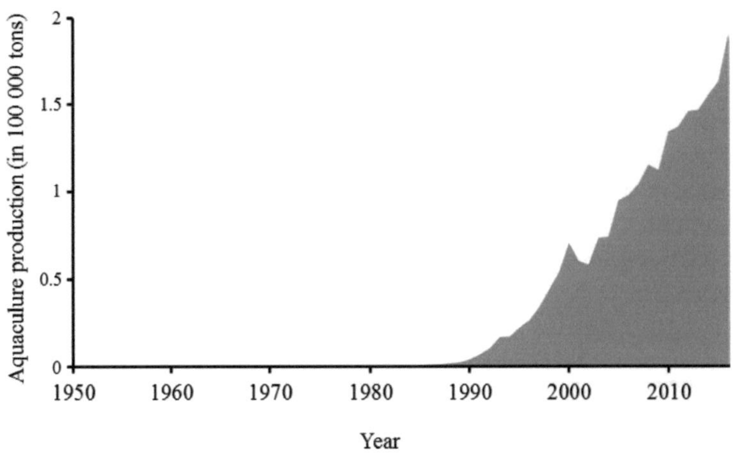

Figure 7. Global aquaculture production from 1950 to 2016 of the seabass *Dicentrarchus labrax* (data from the FAO database).

The seabass has become in less than three decades the second most produced fish species in the Mediterranean Sea, just after the sea bream *Sparus aurata* (whose production has followed a similar trend).

3.3. The Atlantic bluefin tuna (*Thunnus thynnus*)

The farming of the Atlantic bluefin tuna is recent [9]. The first trials of farming date back to the 1970s in Canada, Japan, and Australia [71]. Nevertheless, the aquaculture production truly started in the middle of the 1980s in the Mediterranean Sea, with the evolution of new technics allowing to provide fish to fattening farms [71, 72]. Despite significant progress, notably thanks to the work of a consortium of European researchers, the reliable control of the life cycle of the Atlantic bluefin tuna in captivity was never reached [73, 74]. Consequently, this

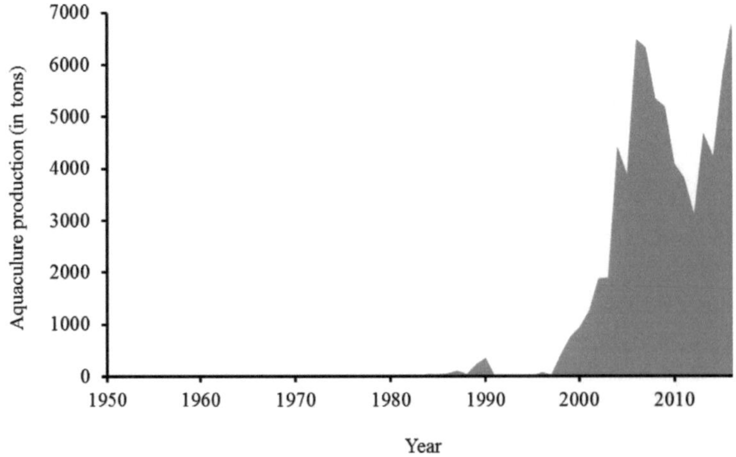

Figure 8. Global aquaculture production from 1950 to 2016 of the Atlantic bluefin tuna *Thunnus thynnus* (data from the FAO database).

industry is entirely based on the stocking of wild-caught specimens, which are reared in cages during a period varying from few months to 2 or 3 years [72–76].

Despite the lack of control of the full life cycle in captivity, the aquaculture production has evolved very quickly from the year 1990 to reach 4080 tons in 2010 (**Figure 8**). More than 10 countries around the Mediterranean Sea currently produce this species, that is, around 60–70 farms [72 - 74]. The leading producers are Croatia, Malta, Turkey, Spain, and Tunisia [77, 78]. The first kind of production, which dates back to the 1980s, is based on the capture of breeders, whose weight varies between 40 and 400 kg, during the migration season, most often close to the spawning areas [73]. Those large individuals are then transported at low speed (1–1.5 knots) over distances ranging from few to 100 km, which might sometimes take several weeks, before being transferred to very large sea cages [73, 79]. Within those very large rearing cages, which may reach 50–60 m of diameter (sometimes even larger than 100 m) and 20–35 m deep, fish are fattened during several months, time required to rebuild muscle fat content that confers to the high-quality flesh researched on the Japanese market of sushi and sashimi [71, 73, 76]. During the middle of 1990s, a second type of production was initiated and consists of capturing immature individuals (8–20 kg in body weight), which are then reared during about 2–3 years in smaller sea cages than for breeders (50–60 m of diameter for about 20 m deep) to get fish of 30–50 kg [73, 76]. This type of production is mainly developed in the Adriatic Sea, particularly in Croatia [71, 76]. In all rearing systems, tunas are most often feed small forage fishes, among which sardinella (*Sardinella aurita*), pilchard (*Sardina pilchardus*), herring (*Clupea harengus*), mackerel (*Scomber scombrus*), chub mackerel (*Scomber japonicas*), horse mackerel (*Trachurus* sp.), bogue (*Boops boops*), and some cephalopods [71, 73, 75, 78, 79]. The feed conversion rates, which are estimated on a wet feed/wet tuna biomass, are most often high and vary between 10 and 20:1 [73, 78]. In other words, it requires 20 kg of forage fish to make 1 kg of tuna [73]. These feed conversion rates may be even beyond 40:1 for large specimens [73]. Two close relative species to the Atlantic bluefin tuna are also currently produced in other geographic areas using similar methods, namely the Northern Pacific bluefin tuna (*Thunnus orientalis*) in Mexico [78, 79] and the Southern bluefin tuna (*Thunnus maccoyii*) in Australia [80]. In 2002, Japanese scientists were able to obtain eggs and larvae of *T. orientalis* in an artificial setting; yet, there is not a broad knowledge of how to culture tuna in captive conditions, and much research is needed to consistently control the entire life cycle in captivity [78].

In conclusion, the aquaculture production of Atlantic bluefin tuna has strongly increased within the past two decades, mainly driven by the Japanese market [78], its high commercial value associated with its high growth (30 kg in 3 years) [73]. However, the production of this species is based on the capture of wild individuals in nature; thus, only a control of the entire life cycle could ensure the sustainability of the industry through a reduction in its reliance on wild stocks [72, 73, 80].

4. Conclusions

The strong increase of the aquaculture production since the early 1980s has relied chiefly on the domestication of an increasing number of fish species [12, 13, 81–85]. Nevertheless, only a limited number has reached a high level of domestication (**Table 1**), such as the rainbow trout, the Nile tilapia, or the Atlantic salmon [83–85]. The 35 species classified at Level 5 [12] belong to 10 families, among which Cyprinidae ($n = 10$ species), Salmonidae ($n = 8$), and Acipenseridae ($n = 5$) [83]. For these species, the entire life cycle is controlled in captivity, and breeding programs have allowed improving, among others, growth, with average genetic gains comprised between 10 and 15% per generation [37, 58, 85–87]. Today, it is estimated that about 10% of the global production is based on improved individuals [37, 87–89]. Nevertheless, very often, even for the species that have reached Level 4 or 5 (**Table 1**), a significant part of global production is based on the introduction of wild individuals. Conversely to these few domesticated species, or more accurately domesticated populations, the majority of farmed fish species still rely on the regular inputs of wild individuals (**Table 1**); thus, there is no strong dichotomy within the same species between wild individuals (coming from fisheries) and farmed individuals (produced in aquaculture) [90–93]. Besides, for numerous species, aquaculture is not a true alternative to capture fisheries but rather a mean to produce wild individuals to a certain commercial size by strongly decreasing the high-mortality rate characteristics of wild populations [90, 94]. Most farmed fish are thus still relatively similar to their wild congeners [95, 96].

Even though the number of farmed aquatic species (including fish, molluscs, and crustaceans) has strongly increased from 1950 to 2010, from about 72 to more than 500 [19, 20], only few species ensure the bulk of the production today [30, 83, 97]. For fish only, 15 species ensure more than 85% of the global production in 2005 [30], despite the number of farmed species rose from 43 to 219 between 1950 and 2005 [97]. In 2009, this trend was confirmed with more than 90% of the global production relying on 20 species only [83]. Only in Europe, most of the aquaculture production is based on the rearing of 10 species only [34, 98]. For some species, which have a very high production today, their farming is quite recent, dating back only to two or three decades only, such as the striped catfish or the Atlantic salmon [30, 97]. Among the 33 species with more than 100,000 tons in 2005, about one-quarter was not produced 40 years ago [97], which illustrates that new species can contribute strongly to the global production [99–101]. Conversely, most farming trials of new species realized within the past decades, either failed or resulted in low production volumes, about tens of tons. This demonstrates how difficult it is to farm a new species, whose development depends on the interaction of various factors, among which biological (availability of wild individuals, ability to control the life cycle in captivity), economical (acceptability by consumers, competition with other animal products), and environmental ones (availability of suitable sites and water, competition with other resources) [12, 18, 84, 91]. More recently, it has also become evident that climate change, which may result, among others, in global warming, saline water intrusion, and ocean acidification, may affect aquaculture [102]. Therefore, aquaculture should use genetically improved and robust animals not suffering from inbreeding depression, resulting from well-managed selective breeding programs with proper inbreeding control and breeding goals [102]. The leading species for aquaculture production have been extensively introduced across the world, particularly in the past century, resulting in that the bulk of aquaculture production relied on the farming of these very few alien species in numerous countries [11, 15]. Yet, the contribution of native species to global aquaculture will perhaps improve resulting in a more diversified and even production than today [99–101]. At least the intensity level of farming, from extensive to highly intensive, and industrialization are also very diverse [18]. From an activity mainly artisanal, aquaculture has evolved to include very large companies that export in numerous countries [17, 103]. According to FAO, the annual production by fish farmer also strongly varies from less than 1 ton in Indonesia to 4 tons in India, 7 tons in China, 35 tons in Chile, and 187 tons in Norway [19].

Conflict of interest

The author declares no conflict of interest.

Author details

Teletchea Fabrice

Address all correspondence to: fabrice.teletchea@univ-lorraine.fr

Université de Lorraine, Inra, URAFPA, Nancy, France

References

[1] Nash CE. The History of Aquaculture. Chichester. Wiley-Blackwell; 2011. 244 p

[2] Cressey D. Future fish. Nature. 2009;**458**:398-400

[3] Harache Y. Development and diversification issues in aquaculture. A historical and dynamic view of fish culture diversification. Cahiers Options Méditerranéennes. 2002;**59**:15-23

[4] Diana JS, Egna HS, Chopin T, Peterson MS, Cao L, Pomeroy R, Verdegem M, Slack WT, Bondad-Reantaso MG, Cabello F. Responsible aquaculture in 2050: Valuing local conditions and human innovations will be key to success. Bioscience. 2013;**63**:255-262

[5] Balon EK. Origin and domestication of the wild carp, *Cyprinus carpio*: From roman gourmets to the swimming flowers. Aquaculture. 1995;**129**:3-48

[6] Liao IC, Huang YS. Methodological approach used for the domestication of potential candidates for aquaculture. Cahiers Options Méditerranéennes. 2000;**47**:97-107

[7] Gjedrem T, Baranski M. Selective breeding in aquaculture: An introduction. In: Reviews: Methods and Technologies in: Fish Biology and Fisheries. Vol. 10. Springer; 2009. 221 p

[8] Pinchot GB. Marine farming. Scientific American. 1970;**223**:15-21

[9] Teletchea F. De la pêche à l'aquaculture. Demain, quels poissons dans nos assiettes? Paris: Editions Belin; Basel. 2016. 180 p. ISBN: 978-2701164397

[10] Gjedrem T, Robinson N, Rye M. The importance of selective breeding in aquaculture to meet future demands for animal protein: A review. Aquaculture. 2012;**350-353**:117-129

[11] De Silva SS, Nguyen TTT, Turchini GM, Amarasinghe US, Abery NW. Alien species in aquaculture and biodiversity: A paradox in food production. Ambio. 2009;**38**:24-28

[12] Teletchea F, Fontaine P. Levels of domestication in fish: Implications for the sustainable future of aquaculture. Fish and Fisheries. 2014;**15**:181-195

[13] Teletchea F. Domestication of marine fish species: Update and perspectives. Journal of Marine Science and Engineering. 2015;**3**:1227-1243

[14] Pauly D, Zeller D. Catch reconstructions reveal that global marine fisheries catches are higher than reported and declining. Nature Communications. 2016;**7**:10244

[15] Shelton WL, Rothbard S. Exotic species in global aquaculture—A review. The Israeli Journal of Aquaculture—Bamidgeh. 2006;**58**:3-28

[16] Diana J. Aquaculture production and biodiversity conservation. Bioscience. 2009;**59**:27-38

[17] Subasinghe R, Soto D, Jia J. Global aquaculture and its role in sustainable development. Reviews in Aquaculture. 2009;**1**:2-9

[18] Bostock J, McAndrew B, Richards R, Jauncey K, Telfer T, Lorenzen K, Little D, Ross L, Handisyde N, Gatward I, Corner R. Aquaculture: Global status and trends. Philosophical Transactions of the Royal Society: B—Biological Sciences. 2010;**365**:2897-2912

[19] FAO. The State of World Fisheries and Aquaculture 2012. Rome: 2012. 209 p

[20] FAO. Contributing to food security and nutrition for all. In: The State of World Fisheries and Aquaculture 2016. Rome: 2016. 200 p

[21] Bartley DM, Nguyen TTT, Halwart M, De Silva SS. Use and exchange of aquatic resources in aquaculture: Information relevant to access and benefit sharing. Reviews in Aquaculture. 2009;**1**:157-162

[22] Jeney Z, Jian Z. Use and exchange of aquatic resources relevant for food and aquaculture: Common carp (*Cyprinus carpio* L.). Reviews in Aquaculture. 2009;**1**:163-173

[23] Balon EK. About the oldest domesticates among fishes. Journal of Fish Biology. 2004; **65**:1-27

[24] Vandeputte M. Selective breeding of quantitative traits in the common carp (*Cyprinus carpio*): A review. Aquatic Living Resources. 2003;**16**:399-407

[25] Vandeputte M. L'amélioration génétique de la carpe commune (*Cyprinus carpio* L.). Cahiers Agriculture. 2009;**18**:256-261

[26] Palaiokostas C, Kocour M, Prchal M, Houston RD. Accuracy of genomic evaluations of juvenile growth rate in common carp (*Cyprinus carpio*) using genotyping by sequencing. Frontiers in Genetics. 2018;**9**. Article 82. DOI: 10.3389/fgene.2018.00082. https://www.frontiersin.org/articles/10.3389/fgene.2018.00082/full

[27] Zheng X, Kuang Y, Lv W, Cao D, Sun Z, Jin W, Sun X. Quantitative trait loci for morphometric traits in multiple families of common carp (*Cyprinus carpio*). Sciences China Life Science. 2017;**60**:287-297

[28] Yuan Z, Huang W, Liu S, Xu P, Dunham R, Liu Z. Historical demography of common carp estimated from individuals collected from various parts of the world using the pairwise sequentially markovian coalescent approach. Genetica. 2018;**146**:235-241

[29] Vitule JRS, Freire CA, Simberloff D. Introduction of non-native freshwater fish species can certainly be bad. Fish and Fisheries. 2009;**10**:98-108

[30] Lazard J, Levêque C. Introductions et transferts d'espèces d'eau douce. Cahiers Agricultures. 2009;**18**:157-163

[31] Koehn JD. Carp (*Cyprinus carpio*) as a powerful invader in Australian waterways. Freshwater Biology. 2004;**49**:882-894

[32] Kestemont P. Different systems of carp production and their impacts on the environment. Aquaculture. 1995;**129**:347-372

[33] Bekefi E, Varadi L. Mutlifunctional pond fish farms in Hungary. Aquaculture International. 2001;**15**:227-233

[34] Varadi L. Review of trends in the development of European inland aquaculture linkages with fisheries. Fisheries Management and Ecology. 2001;**8**:453-462

[35] Réalis-Doyelle E, Pasquet A, Fontaine P, Teletchea F. Effects of temperature on the survival rate and development during the endogenous feeding period of common carp (*Cyprinus carpio*). Hydrobiologia. 2018;**805**:365-375

[36] Lazard J. La pisciculture des tilapias. Cahiers Agricultures. 2009;**18**:174-182

[37] Gjedrem T. Genetic improvement for the development of efficient global aquaculture: A personal opinion review. Aquaculture. 2012;**344-349**:12-22

[38] Bentsen HB, Gjerde B, Eknath AE, Palada de Vera MS, Velasco RR, Danting JC, Dionisio EE, Longalong FM, Reyes RA, Abella TA, Tayamen MM, Ponzoni RW. Genetic improvement of farmed tilapias: Response to five generations of selection for increased body weight at harvest in *Oreochromis niloticus* and the further impact of the project. Aquaculture. 2017;**468**:206-217

[39] Sukmanomon S, Kamonrat W, Poompuang S, Nguyen TTT, Bartley DM, May B, Na-Nakorn U. Genetic changes, intra-and inter-specific introgression in farmed Nile tilapia (*Oreochromis niloticus*) in Thailand. Aquaculture. 2012;**324-325**:44-54

[40] Thodesen J, Rye M, Wang Y-X, Bentsen HB, Gjedrem T. Genetic improvement of tilapias in China: Genetic parameters and selection responses in fillet traits of Nile tilapia (*Oreochromis niloticus*) after six generations of multi-trait selection for growth and fillet yield. Aquaculture. 2012;**366-367**:67-75

[41] Gall GAE, Crandell PA. The rainbow trout. Aquaculture. 1992;**100**:1-10

[42] Fornshell G. Rainbow trout—Challenges and solutions. Reviews in Fisheries Science. 2002;**10**:545-557

[43] Vandeputte M, Baroillier J-F, Hafrray P, Quillet E. Amélioration génétique des poissons: quelles réalisations et quels défis pour demain? Cahiers Agriculture. 2009;**18**:262-269

[44] Janssen K, Chavanne H, Berentsen P, Komen H. Rainbow Trout (*Oncorhynchus mykiss*)—Current Status of Selective Breeding in Europe. 2015. Pdf available on line [Accessed: May 15, 2018]

[45] Janssen K, Chavanne H, Berentsen P, Komen H. Impact of selective breeding on European aquaculture. Aquaculture. 2017;**472**:8-16

[46] Adeli A, Baghaei F. Production and supply of rainbow trout in Iran and the world. World Journal of Fish and Marine Sciences. 2013;**5**:335-341

[47] Solar II. Use and exchange of salmonid genetic resources relevant for food and aquaculture. Reviews in Aquaculture. 2009;**1**:174-196

[48] Cacot P, Lazard J. La domestication des poissons du Mékong: les enjeux et le potentiel aquacole. Cahiers d'Agriculture. 2009;**18**:125-135

[49] Lazard J, Cacot P, Slembrouk J, Legendre M. La pisciculture des Pangasiidae. Cahiers Agriculture. 2009;**18**:164-173

[50] Nguyen TTT. Patterns of use and exchange of genetic resources of the striped catfish *Pangasianodon hypophthalmus* (Sauvage 1878). Reviews in Aquaculture. 2009;**1**:224-231

[51] Bui TM, Phan LT, Ingram BA, Nguyen TTT, Gooley GJ, Nguyen HV, Nguyen PT, De Silva SS. Seed production practices of striped catfish, *Pangasianodon hypophthalmus* in the Mekong Delta region, Vietnam. Aquaculture. 2010;**306**:92-100

[52] Phan LT, Bui TM, Nguyen TTT, Gooley GJ, Ingram BA, Nguyen HV, Nguyen PT, De Silva SS. Current status of farming practices of striped catfish, *Pangasianodon hypophthalmus* in the Mekong Delta, Vietnam. Aquaculture. 2009;**296**:227-236

[53] Tacon AGJ, Metian M, Turchini GM, De Silva SS. Responsible aquaculture and trophic level implications to global fish supply. Reviews in Fisheries Science. 2010;**18**:94-105

[54] De Silva SS. Aquaculture: A newly emergent food production sector—And perspectives of its impacts on biodiversity and conservation. Biodiversity and Conservation. 2012; **21**:3187-3220

[55] Subasinghe R. Regional review on status and trends in aquaculture development in Asia-Pacific—2015. FAO Fisheries and Aquaculture Circular No. 1135/5. 2007. ISSN: 2070-6065

[56] Murk AJ, Rietjens IMCM, Bush SR. Perceived versus real toxicological safety of pangasius catfish: A review modifying market perspectives. Reviews in Aquaculture. 2018;**10**: 123-134

[57] Gjedrem T. The first family-based breeding program in aquaculture. Reviews in Aquaculture. 2010;**2**:2-15

[58] Gjedrem T. Genetic improvement of cold-water fish species. Aquaculture Research. 2000; **31**:25-33

[59] Gutierrez AP, Yáñez JM, Davidson WS. Evidence of recent signatures of selection during domestication in an Atlantic salmon population. Marine Genomics. 2016;**26**:41-50

[60] Glover KA, Solberg M, McGinnity P, Hindar K, Verspoor E, Coulson MW, Hansen MM, Araki H, Skaala Ø, Svåsand T. Half a century of genetic interaction between farmed and wild Atlantic salmon: Status of knowledge and unanswered questions. Fish and Fisheries. 2017;**18**:890-927

[61] Tacon AGJ, Halwart M. Cage aquaculture: A global overview. In: Halwart M, Soto D, Arthur JR, editors. Cage Aquaculture—Regional Reviews and Global Overview. FAO Fisheries Technical Paper. No. 498 (pp. 1-16). FAO: Rome; 2007. 240 p

[62] Niklitschek EJ, Soto D, Lafon A, Molinet C, Toledo P. Southward expansion of the Chilean salmon industry in the Patagonian fjords: Main environmental challenges. Reviews in Aquaculture. 2013;**5**:172-195

[63] Carvajal J, González L, George-Nascimento M. Native sea lice (Copepoda: Caligidae) infestation of salmonids reared in netpen systems in southern Chile. Aquaculture. 1998; **166**:241-246

[64] Barton JR, Fløysan A. The political ecology of Chilean salmon aquaculture, 1982-2010: A trajectory from economic development to global sustainability. Global Environmental Change. 2010;**20**:739-752

[65] Badiola M, Mendiola D, Bostock J. Recirculating aquaculture systems (RAS) analysis: Main issues on management and future challenges. Aquacultural Engineering. 2012;**51**: 26-35

[66] Chavanne H, Chatain B, Haffray P. European seabass (*Dicentrarchus labrax* L.). Review on breeding and reproduction of European aquaculture species. Aquabreeding. 2008

[67] Chavanne H, Janssen K, Hofherr J, Contini F, Haffray P, Aquatrace Consortium, Komen H, Nielsen EE, Bargelloni L. A comprehensive survey on selective breeding programs and seed market in the European aquaculture fish industry. Aquaculture International. 2016;**24**:1287-1307. https://link.springer.com/article/10.1007/s10499-016-9985-0)

[68] Chatain B, Chavanne H. La génétique du bar (*Dicentrarchus labrax* L.). Cahiers Agricultures. 2009;**18**:249-255

[69] Novel P, Porta J, Fernández J, Méndez T, Gallardo-Gálvez JB, Béjar J, Alvarez MC. Critical points for the maintenance of genetic variability over a production cycle in the European sea bass. *Dicentrarchus labrax*. Aquaculture. 2013;**416-417**:8-14

[70] Vandeputte M, Dupont-Nivet M, Haffray P, Chavanne H, Cenadelli S, Parati K, Vidal M-O, Vergnet A, Chatain B. Response to domestication and selection for growth in the European sea bass (*Dicentrachus labrax*) in separate and mixed tanks. Aquaculture. 2009;**286**:20-27

[71] Fauvel C, Suquet M. La domestication des poissons: le cas du thon rouge. INRA Productions Animales. 2004;**17**:183-187

[72] Milatou N, Megalofonou P. Age structure and growth of bluefin tuna (*Thunnus thynnus* L.) in the capture-based aquaculture in the Mediterranean Sea. Aquaculture. 2014;**424-425**: 35-44

[73] Mylonas CC, De la Gándara F, Corriero A, Belomonde RA. Atlantic bluefin tuna (*Thunnus thynnus*) farming and fattening in the Mediterranean Sea. Reviews in Fisheries Science. 2010;**18**:266-280

[74] Reglero P, Balbín R, Ortega A, Alvarez-Berastegui D, Gordoa A, Torres AP, Moltó V, Pascual A, De la Gándara F, Alemany F. First attempt to assess the viability of bluefin tuna spawning events in offshore cages located in a priori favourable habitat. Scientia Marina. 2013;**77**:585-594

[75] Sylvia P, Belle S, Smart A. Current status and future prospective of bluefin tuna (*Thunnus thynnus orientalis*) farming in Mexico and the West Coast of the United States. In: Bridges CR, García A, Gordin H, editors. Domestication of the Bluefin Tuna *Thunnus thynnus thynnus*. 60. Cahiers Options Méditerranéennes. Zaragoza: CIHEAM; 2003. pp. 197-200

[76] Katavić I, Tičina V. Fishing and farming of the northern bluefin tuna (*Thunnus thynnus* L.) in the Adriatic Sea. In: Cataudella S, Massa F, Crosetti D, editors. Interactions between Aquaculture and Capture Fisheries. A Methodological Perspective. Vol. 78. Studies and Reviews. General Fisheries Commission for the Mediterranean. Rome. 2005. 229 p. www.fao.org/docrep/009/a0141e/a0141e00.htm

[77] Muňoz-Lechuga R, Sanz-Fernández V, Cabrera-Castro R. An overview of freshwater and marine finfish aquaculture in Spain: Emphasis on regions. Reviews in Fisheries Science & Aquaculture. 2018;**26**:195-213

[78] Loo M. Tuna Farming in Mexico: Is their a Sustainable Future? Ecological Aquaculture Studies and Reviews. Kingston, RI: University of Rhode Island; 2007

[79] Zertuche-González JA, Sosa-Nishizaki O, Vaca Rodriguez JG, del Moral SR, Yarish C, Costa-Pierce BA. Marine science assessment of capture-based tuna (*Thunnus orientalis*) aquaculture in the Ensenada Region of Northern Baja California, Mexico. Final Report to the David & Lucile Packard Foundation, Los Altos, California, USA. 2008. 95 p

[80] Bubner E, Farley J, Thomas P, Bolton T, Elizur A. Assessment of reproductive maturation of southern bluefin tuna (*Thunnus maccoyii*) in captivity. Aquaculture. 2012;**364-365**:82-95

[81] Hong W, Zhang Q. Review of captive bred species and fry production of marine fish in China. Aquaculture. 2003;**227**:305-318

[82] Teletchea F. Qu'est-ce qu'un poisson domestique? Implications pour le développement futur de l'aquaculture. Ethnozootechnie. 2012;**90**:7-12

[83] Teletchea F. Fish domestication in aquaculture: Reassessment and emerging questions. Cybium. (submitted)

[84] Bilio M. Plenary lecture: The future of capture and culture fisheries—Keeping and creating diversity in fish production. In: Arai K, editor. Entitled Innovative Marine Life Science for Three Es, Edibles Environment and Education, in 21st Century (Proceedings of the 7th International Symposium, 17-19 November 2008). Hokkaido, Japan. 2008. pp. 5-24. https://www.yumpu.com/en/document/view/29619809/plenary-lecture-the-future-of-capture-and-culture-fisheries

[85] Hedgecock D. Aquaculture, the next wave of domestication. In: Gepts P, Famula TR, Bettinger RL, Brush SB, Damania AB, McGuire PE, Qualset CO, editors. Biodiversity in Agriculture, Domestication, Evolution, and Sustainability. Cambridge. Cambridge University Press; 2012. pp. 538-548

[86] Nguyen NH. Genetic improvement for important farmed aquaculture species with a reference to carp, tilapia and prawns in Asia: Achievements, lessons and challenges. Fish and Fisheries. 2016;**17**:483-506

[87] Gjedrem T, Robinson N. Advances by selective breeding for aquatic species: A review. Agricultural Sciences. 2014;**5**:1152-1158

[88] Olesen I, Betnsen HB, Philipps M, Ponzoni RW. Can the global adoption of genetically improved farmed fish increase beyong 10%, and how? Journal of Marine Science and Engineering. 2015;**3**:240-266

[89] Gjedrem T, Rye M. Selection response in fish and shellfish: A review. Reviews in Aquaculture. 2018;**18**:168-179

[90] Ottolenghi F, Silvestri C, Giordano P, Lovatelli A, New MB. Capture-based Aquaculture. The Fattening of Eels, Groupers, Tunas and Yellowtails. Rome: FAO; 2004. 308 p

[91] Klinger DH, Turnipseed M, Anderson JL, Asche F, Crowder LB, Guttormsen AG, Halpern BS, O'Connor MI, Sagarin R, Selkoe KA, Shester GG, Smith MD, Tyedmers P. Moving beyond the fished and farmed dichotomy. Marine Policy. 2013;**38**:369-374

[92] Teletchea F. Is fish domestication going too fast? Natural Resources. 2016;**7**:399-404

[93] Teletchea F. Chapter 20: Domestication and genetics: What a comparison between land and aquatic species can bring? In: Pontarotti P, editor. Evolutionary Biology: Biodiversification from Genotype to Phenotype. Basel. Springer; 2015. pp. 389-401

[94] Naylor RL, Goldburg RJ, Primavera JH, Kautsky N, Beveridge MCM, Clay J, Folke C, Lubchenco J, Mooney H, Troell M. Effect of aquaculture on world fish supplies. Nature. 2002;**405**:1017-1024

[95] Gjerde B, Rye M. Design of breeding programmes in aquaculture species: Possibilities and constraints. In: Bartley DM, Basurco B, editors. Genetics and Breeding of Mediterranean Aquaculture Species Zaragoza. Vol. 34. CIHEAM. Cahiers Options Méditerranéennes. 1998. pp. 181-192. http://om.ciheam.org/option.php?IDOM=768

[96] Teletchea F, Fontaine P. Beyond the levels of domestication in fish: Must all farmed species be domesticated? In: Proceedings of the VI International Conference "Water and Fish" June 12-14, University of Belgrade, Serbia. 2013. pp. 74-81

[97] Fontaine P, Legendre M, Vandeputte M, Fostier A. Domestication de nouvelles espèces et développement durable de la pisciculture. Cahiers Agricultures. 2009;**18**:119-124

[98] Turchini GM, De Silva SS. Bio-economical and ethical impacts of alien finfish culture in European inland waters. Aquaculture International. 2008;**16**:243-272

[99] Sicuro B, Luzzana U. The state of *Seriola* spp. other than yellowtail (*S. quinqueradiata*) farming in the world. Reviews in Fisheries Science & Aquaculture. 2016;**24**:314-325

[100] Rimmer MA, Glamuzina B. A review of grouper (Family Serranidae: Subfamily Epine phelinae) aquaculture from a sustainability science perspective. Reviews in Aquaculture. 2017. https://onlinelibrary.wiley.com/doi/abs/10.1111/raq.12226

[101] Ramos Valladão GM, Gallani SU, Pilarski F. South American fish for continental aquaculture. Reviews in Aquaculture. 2016;**10**:351-369. https://onlinelibrary.wiley.com/doi/abs/10.1111/raq.12164

[102] Sae-Lim P, Kause A, Mulder HA, Olesen I. Breeding and genetics symposium: Climate change and selective breeding in aquaculture. Journal of Animal Sciences. 2017;**95**: 1801-1812

[103] Longo SB, Clark B, York R. The globalization of ecologically intensive aquaculture (1984-2008). Journal of Environmental Studies and Science. 2013;**3**:297-305

Chapter 5

Effects of Domestication on Fish Behaviour

Alain Pasquet

Additional information is available at the end of the chapter

http://dx.doi.org/10.5772/intechopen.78752

Abstract

Domestication is a process by which humans select some phenotypes of wild animal species (i.e., morphological traits or growth), but as all traits are linked, the selection of a particular one has consequences on others. In that context, behavioural traits may be affected by human selection. In this chapter, through classical behavioural traits, such as swimming capacities, foraging, social interactions, or reproduction, and also personality or cognitive abilities, what domestication modifies in fish behavioural traits is shown. The information is taken only from studies that make a clear comparison between domesticated and wild animals; the major difficulty was that the domesticated status was not clearly determined. Whatever the behavioural trait considered, domestication affects some of them even after only one generation. These data deserve to be taken into consideration when humans try, not only to domesticate new species but also to release domesticated species into their natural habitats. In this last case, alteration of behavioural traits could make the fish incapable to adapt to their new wild environment and alter their foraging or reproductive performances. Moreover, fish behaviour in farm is currently recognised as an essential component of the welfare and all behavioural modifications must be considered.

Keywords: behavioural traits, hatchery-reared fish, wild fish, performances, behavioural responses

1. Introduction

Behaviour is an animal phenotype and could be considered as a variable of adjustment for an animal to changes of environmental factors. Domestication gives new environmental conditions to animals; they have to adapt to these restricted surroundings. In general, captive conditions are less complex than those of a natural environment but even with less complexity, the environmental conditions of farms or other rearing structures could appear as new for animals. So they have to adapt. As a phenotype, behaviour is certainly the mean and the most useful to survive under the new conditions. So during the domestication process, behaviour allows the animal to adapt to its new environmental conditions. Through domestication, the artificial selection is a process of changing characteristics of animals by artificial means such as directional selection, familial selection [1] or genomic selection [2], and the domestication may impact the behaviour even after only one generation [3, 4].

Behavioural traits are among the first traits to be affected by domestication [5, 6]. Behaviour is more easily moulded than morphology or chemical composition and thus the costs of behavioural modification are more efficiently adjusted to environmental variations. In his book, Jensen [7] described the effects of domestication in vertebrates, mainly on birds and mammals but there was nothing on fishes. Before that, there were three major reviews [3, 4, 8] on the influence of aquaculture and domestication on fish behaviour. In these papers, the authors summarised most of the available information on the effects of domestication on different traits of fish behaviour. The major aim of these reviews was to consider the importance of behavioural modifications due to domestication on the economic interest of the culture of fishes and on the welfare of animals in fish farms. In this chapter, I focus on the behavioural traits that have been modified by domestication without consideration to either economic objectives or animal welfare.

There are many difficulties to analyse papers dealing with the effects on domestication. First, it is not easy to identify precisely neither the number of generations in captivity nor the link between captive and wild animals. It is easy when it concerns the first generation obtained in captivity, but it is more complex when we address to 'individuals reared in hatcheries' for several years. Most often, we do not know if there was time introduction of wild animal (e.g. males) during the domestication process. Second, in most studies comparing wild and domesticated strains, we have very few information on the characteristics of the wild animals and on those of their native sites. It is important because there is an important variability of the behavioural trait parameters between different populations. Third, in general, fish performances of behavioural traits are tested under laboratory conditions except for displacements for which some experiments were realised in natural water areas. So whatever the experimental sites, the foreigner population (wild or domesticated) needs a period of acclimation to its new rearing conditions. These could introduce a bias in the results.

Behaviour is the basis of all relationships between the animal and its environment and concerns with several behavioural traits: swimming, foraging, predator avoidance, relationships with conspecifics and reproduction. Moreover, it is now known that individuals exhibit behavioural or physiological characteristics, which, if they are consistent over time, define a coping style or personality [9]. As through domestication, human beings select some individuals among a population, this could modify the equilibrium between the different behavioural profiles (or coping styles) of the individuals of a population. Now, some researches integrate this individual component and highlight the effects of domestication on individual behaviour as it has recently been done considering the learning and other cognitive abilities of fish.

In this chapter, I will review some of these behavioural traits in hatchery-reared fishes that have often been altered in a characteristic manner by domestication.

2. Swimming behaviour

Swimming is a general behavioural trait, which is used in different situations: foraging activity, predator avoidance, stress responses or reproduction. For fish, one of the most determinant traits that are able to improve foraging is the swimming ability. In rearing conditions, swimming is no longer as important as in nature; in general, fish have less space at their disposal, but if domestication selects individuals on their morphological and physiological characteristics, this could influence directly their swimming performances.

This behaviour trait has been tested on fishes in response to a predator attack. It is the case for juveniles (between 55 and 125 days old) of the sea bass (*Dicentrarchus labrax*); wild individuals showed a greater angular velocity and a stop distance to a new object more important than reared fishes [10]. These responses decrease with habituation in both groups. It means that wild individuals have a greater reactivity and a longer escape distance from an unknown object in their environment.

In the context of swimming behaviour, one of the more common tested parameter is the C-start response: this is the ability of an individual to rest from a novel environmental situation; it is characterised by a rapid reaction of the body with a C posture and after an S followed by a rapid (less than 10 ms) displacement. It measures the physical ability of a fish to react to a stress situation by using its physical abilities to swim. It has been tested in different environmental situations: pollution [11], water temperature [12], hypoxia [13] or the influence of conspecific presence by comparing solitary and grouped individuals [14]. In all cases, wild fishes showed a greater velocity and more rapid swimming abilities, so it seems that domestication decreases the swimming performances of the fish. This decrease could be parallel to physiological events. Comparisons of swimming and metabolic physiology were done in aquaculture-reared California yellowtail (*Seriola dorsalis*) in comparison to wild individuals. Incremental swimming velocity trials showed that aquaculture-reared fish had a significantly slower mean maximum sustainable swimming speed (4.16 ± 0.62 Body Length s^{-1}) in comparison to that of wild fish (4.80 ± 0.52 BL s^{-1}). In addition, oxygen consumption was significantly higher in aquaculture-reared fish (7.31 ± 2.32 vs. 3.94 ± 1.60 mg O_2 kg^{-1} min^{-1} at 18°C) in comparison to wild-caught yellowtail (15.80 ± 5.78 mg O_2 kg^{-1} min^{-1}) [15].

This could alter other behaviours, which depend directly on swimming (i.e. foraging, survival). One point that concerns with swimming performances is the ability for reared individuals to be released in wild sites. This is the case for the European grayling (*Thymallus thymallus*) that were tagged with radio-transmitters and tracked in the Blanice River, River Elbe catchment (Czech Republic) [16]. Wild and hatchery-reared fish increased their dial movements and home range with environmental variables (light intensity, flow, temperature and turbidity), but hatchery-reared fish displayed greater total migration distance than did wild fish, which was caused mainly by their higher dispersal. Patterns in space use and activity were compared for wild and hatchery-reared Mulloway (*Argyrosomus japonicus*) using acoustic telemetry. Adult individuals were followed during 288 h in a river. Hatchery-reared fish used significantly larger areas with higher rate of activity than wild fish, but their movement ranges were more variable [17] than those of wild fish. By comparing initial movement, habitat use, growth and mortality between stocked hatchery and wild fish of juveniles of Florida Bass (*Micropterus floridanus*) with a radio telemetry experiment, Thomson et al. [18] showed that tagged hatchery fish exhibited greater movement (75 and 124 m/d, respectively), greater proportion of locations offshore (8 and 23%, respectively), but slower growth (1.73 and 0.41% of their body weight gained per day, respectively), and higher predation (47 and 0%, respectively) than wild fish.

These results showed that domestication can not only be influenced through selecting the physical characteristics of the individuals, but also through their swimming performances and consequently the foraging and space use by hatchery-reared individuals when released in wild conditions.

3. Foraging behaviour

Foraging is not only the activity, which consists to take off resources in the environment, that is, prey, but also the choice of the best site or the most favourable period where and when to forage. The animal must be at the good place at the best moment. This aim seems easy for animals in controlled environments where the food is abundant and regular; but this fact could be a disadvantage when aquaculture-reared fish are released in natural environment in order to supply the low level of the wild stocks.

Fishes change their foraging habits with domestication. Zebra fish (*Danio rerio*) and coho salmon (*Oncorhynchus kisutch*) change the place where they forage after domestication after just one generation. Domestic fishes swim at the surface of the water column instead of the lower part for wild animals [19, 20]. One of the consequences is that farmed animals had a higher rate of prey capture than their wild congeners [21, 22]. These changes in foraging behaviour could be the result of changes in the relation of the fishes with its environment: as the predation rate was lower for farmed fishes, they adopt a more risky behaviour near the surface; the farmed conditions modified also the social relationships between individuals and could result in a lower influence of dominance in the foraging behaviour [23].

Perhaps, the main difference is that the natural environment provides a lot of different situations to which fishes have to adapt. It seems that the environmental complexity of natural environments may facilitate training to different situations [24], with a more important prey variability [25–27] or opportunity of social learning [28]. Consequences could be measured when farmed fishes were realised into natural environment: they use less of natural objects such as stones or leaves for digestion than wild animals [25] or they make no difference between prey of different profitability [26] and they do not choice an unknown prey [27].

The conditions of foraging allow the fish to get a certain amount of resources from the environment and could explain important differences between hatchery-reared and wild individuals in terms of survival and growth. If we compare the survival rate of aquaculture-reared or wild Chinook salmon fry (*Oncorhynchus tshawytscha*) facing predation by rainbow trout (*Oncorhynchus mykiss*) or sculpin (*Cottus rhotheus*) under experimental conditions, wild fry had a survival advantage within the two next years of experiment [29]. So it is possible that the domestication can affect the vulnerability of juveniles of salmon after only one generation in a culture system. But it is not always the case. For example, the survival of Atlantic salmon (*Salmo salar)* in the Baltic Sea was examined in relation to the origin, and prey fish abundance (here herring *Clupea harengus* and sprat *Sprattus sprattus*). The study was based on recapture data for tagged hatchery-reared, and wild smolts demonstrated a combined influence of origin and environmental factors on survival; prey fish abundance had no influence on the survival of reared or wild smolt groups [30]. The results suggest that some larger smolt of the reared groups compared with the wild groups compensated for their lower ability to live in the wild.

4. Predator avoidance behaviour

The anti-predator behaviour is highly sensitive to artificial rearing and so to domestication [12, 31–36]. Anti-predator behaviour is thought to change during domestication, along with other traits. One prediction is that domestication should reduce behavioural responses to predation risk. This prediction was supported by a lot of studies most of the time on salmonids, on rainbow trout (*Oncorhynchus mykiss*) [31, 32], on brown trout (*Salmo trutta***)** [12] and on Atlantic salmon (*Salmo salar*) [35, 37].

In wild population, decreased activity, spatial avoidance of risky areas and the use of refuges reduce the rate of mortality caused by predators [38, 39]. This natural reaction of a fish faced to a high level of predation seems to disappear after two or three generations reared under artificial conditions; that is, after two generations, the common trout becomes non-sensitive to the predation risk; animals were active during the daylight and not during the night as their wild conspecifics [40]. As a consequence, domestication would decrease the level of defences against predators, as the reared animals would not experiment contacts with predators or some other life history traits should be affected by domestication and consequently affect the response of the animal to predator risk. For example, wild fishes react more rapidly to a predator than reared fishes [41, 42]. Wild animals may use natural refuges in their environment they know to escape from predation [43]. Moreover, wild individuals seem more careful to predators than reared fishes in the common carp (*Cyprinus carpio*); but these results are under suspicion because 'wild' animals are in fact reared individuals, which were returned back to natural conditions [44]. Domestication may also affect the reaction to a novel object in the environment; reared fishes approach more easily to a novel object and take more risks [36, 45]. This difference in behaviour is linked to physiological variations (heart activity, mobility, swimming abilities…) [35, 37]: but the results are not so clear and in a large number of cases, the responses of reared fishes to predators are variable [19, 46].

Some more recent results confirm the complexity of the relationships between this behavioural trait (anti-predator behaviour) and domestication. For example, the anti-predator behaviour of juvenile Atlantic salmon of conventional hatchery compared with that of wild-caught juveniles from the same population, tested in two unfamiliar environments, did not differ between the two strains in the spontaneous escape response [47], but after this first reaction, hatchery-reared juveniles stayed less time in association with the shelter than the wild animals. The same result has been found in the grass carp (*Ctenopharyngodon idella*); in the frame of restocking programs using hatchery-reared individuals, it is important to test the anti-predator behaviour. This behaviour was compared with that of wild-caught animals. The two groups exhibited a clear anti-predator behaviour; however, the hatchery-reared individuals showed lower aggregation and spent time in the risky areas and most of them were predated [48]. These variations between domesticated and wild strains in the display of the anti-predator behaviour are well documented in rainbow trout (*Oncorhynchus mykiss*). Comparisons between wild and hatchery population between clonal lines of rainbow trout derived from either wild and hatchery-reared populations identified several genes associated with behavioural variations between lines [49]. These genetic variations underlying anti-predator behaviours may be used in conservation programs for monitoring alleles of loci affecting predation in natural populations.

As behaviour is a phenotype corresponding to the plasticity of the responses of animal to the set of environmental conditions, it is interesting to understand how development can affect the behaviour of different genotypes. Now, the existence of transgenic species offers a good tool to study this problem. By comparing wild-type siblings and transgenic individuals, Sundström et al. [50] found that wild and transgenic animals behave in the manner under natural like conditions; but until now, there are not a sufficient number of studies to conclude that genetically modified organisms are not affected by the complexity of natural conditions.

5. Social behaviour

Social behaviour is particularly developed in fishes, such as shoal [51], which is a part of the social life and is present in more than 25,000 species [52]. Shoal is important and ensures protection against a potential predator (a particular prey is undetectable in the group), but also it increases the foraging efficiency (the amount of food per individual is higher in groups than for solitary fishes whatever their diet). Shoal—defined as a group of individuals [51]—may be influenced by environmental factors, and domestication is one of these factors; reared conditions modify the fish environment. It limits the available space for fishes that could have for consequences a non-response of the fishes to environmental stimuli [53]; in reared conditions, food is distributed *ad libitum*, and such situation modifies the foraging behaviour limiting the exploration of the environment [54] and the predator avoidance [12, 32, 55]. In domesticated fishes, there is less variability of the age and size of the individuals, and so, the relations between fishes are modified and the results are counterbalanced; in some studies, they show that there is an increase of the aggressiveness between individuals [56, 57], and in other studies, they find that the aggressiveness is higher in domesticated populations [55, 58]. Growth in rearing situations is influenced by intra-specific competition [59, 60].

One of the most important components of the social relations between individuals is the agonistic behaviour. Comparisons between wild and reared fishes show that new agonistic behaviours do not appear due to domestication [61]; agonistic behaviours are the same for both wild-reared individuals. In general, agonistic behaviours appear for the competition for resources: prediction is that agonistic behaviours must be less numerous when the quantity of resources increases. Domestication introduces the selection of individuals with a rapid growth; the consequences on the level of agonistic behaviours between individuals inside the groups are very dependent of the situation. Globally, it has been demonstrated that an effect on agonistic behaviours exists [62]. Agonistic behaviour can increase for domesticated fishes [58, 63, 64] or decrease [56] or be stable [57]. For example, the brown trout sea-ranched individuals have a higher growth rate and have no difference of activity with wild animals, but intensity of agonistic behaviours was higher in wild individuals [65]. These results could be interpreted as a consequence of the rearing conditions; in wild populations, agonistic behaviour has a function for space sharing, food accessibility [66], foraging efficiency and predator avoidance [67, 68]. So selection in rearing conditions leads to the individuals that have the most rapid growth but with particular behavioural traits (i.e. the most aggressive fishes); it is a known phenomenon, analysed as phenotypic selection (or economic selection by culturists) [69]. This implies that fishes are selected on their size and growth rate, and the dominance effect, which could be the result of competitive relationships, disappears if we introduce the size as variable [23]. But the dominance depends on the environment; this could be linked to the residence effect, which exists in wild fishes and not in reared ones [70]. In any case, competitive behaviours are the same; they vary in quality and intensity between wild and reared fishes [71]; for example, the high density for reared fishes in tanks could induce less territoriality and so a lower aggressiveness during dyadic confrontations [70, 72]. Competition and dominance have been tested in the salmon (*Oncorhynchus tshawytscha*) and the results showed that wild fishes were more aggressive than fishes from the first generation (F1) reared in aquaculture [73]. In general, the consequence of dominance is better growth rates for the dominant individuals whatever their origin (wild or reared). More recently, a relationship was found on the influence of domestication on brain size and aggressive behavioural changes. A study on rainbow trout lines highlighted that some behaviours such as 'freeze' and 'escape' are associated with a high level of domestication instead of 'display' and 'yawn' behaviours, which are linked to wild lines [74]. Moreover, these authors found that the total brain size and olfactory volume were associated with domestication.

An important consequence of the level of aggressiveness between individuals is the existence of cannibalism [75]. It could appear either within the same cohort or between different cohorts. Cannibalism is a natural phenomenon, which is for regulating natural populations in many fish species. In cultured fishes, cannibalism has a negative effect on the populations; some individuals switch from food given by humans to the attacks and consumption of conspecifics.

6. Reproduction

There is very few data on the influence of domestication or different lineages on the reproductive behaviour of fishes? This is the consequence that the reproductive behaviour in reared fishes received very little interest. It is the consequence that humans biased reproduction in reared fish populations; in fact, it is always handed by humans, and there is neither mate choice nor normal reproductive behavioural sequence. So, comparisons of reproductive behaviours between wild and reared fishes are based on behavioural differences between reared fishes that returned to natural environment and wild animals.

Studies focused on the choice of the spawn area; reared animals had more difficulties to find the good place to spawn with environmental features [76]. But the results are not so clear. Reared fishes may arrive earlier on the spawning zones than wild animals [77]. Fishes show different strategies with regard to their origin (wild or reared) [30, 78].

Most of the studies on the influence of domestication on the reproductive behaviour are done on salmonids because this is the group of species with the highest pressure for restocking the natural populations with hatchery-reared individuals, so it is absolutely necessary to evaluate their reproductive performances under natural environment. Coho Salmon (*Oncorhynchus kisutch*) produced by hatcheries have lower fitness in the wild than naturally produced salmon, but the factors underlying this difference remain an active area of research [71]. Neff et al. [79] used genetic parentage analysis of juveniles produced by experimentally mixed groups of wild and hatchery coho salmon to quantify male paternity. In all contexts, wild animals showed a higher paternity rate than hatchery-reared individuals.

7. Personality

The concept of behavioural syndrome (synonyms = personality, temperament, behavioural differences) is defined as a collection of behavioural traits, which are constant over time and environmental situations [80]. It does not mean that these traits do not evolve with time for example, but that the combination of them is constant. This concept has been widely used in fishes. These behavioural syndromes may be dependent from the environmental situations (i.e. high or low density) and have different performances (i.e. boldness or shyness are the most efficient). This concept has been used for cultured fishes (*Salmonidae*) in order to select the most advantageous behavioural traits for the rearing of fishes in captivity. The human selection on economic criteria (size, growth) may be biased and this selection leads to keep the individuals that have the highest boldness (as in *Salmonidae*). But these results are not so clear, and in some cases, the selection of the individuals, which have the highest boldness, leads also to the selection of the most aggressive animals, i.e. salmon reared in farm for many generations are more aggressive and bold than individuals hatched in farm but from wild parents [72, 81]. Now, it is possible by comparing wild and domesticated strains, to show the existence of QTL for personality trait such as boldness. By testing the boldness of Zebra fish (*Danio rerio*), Wright et al. [54] showed that there are strong behavioural differences between a wild-derived strain of fish and a laboratory strain AB. Based on anti-predator behaviour, their results indicated a QTL for boldness on chromosomes 9 and 16 and suggest another genomic region that influences anti-predator behaviour on chromosome 21. So, these results confirm the possibility of QTL mapping of behavioural traits in zebra fish and the consequences of selection during domestication.

These behavioural differences between captive of reared fish and their wild conspecifics could be used in the frame recovery programmes for threatened and endangered species. By comparing the boldness and prey acquisition behaviours of wild bull trout (*Salvelinus confluentus*) and reared ones, Brignon et al. [82] showed that wild fish and captive reared fish from complex habitats exhibited a greater level of boldness and prey acquisition ability, than fish reared in conventional captive environments. These results suggested that rearing fish in more complex captive environments could create a more wild-like phenotype than conventional rearing practices.

In this frame of animal personality, or coping style, an important effect of the domestication is the reduction of emotional reactivity or responsiveness to a fear-evoking stimulus [83]; the emotional reactivity of wild fishes is better than those of reared individuals [84]. The emotional reactivity of an animal is necessary for provoking a flight response when there is a potential danger; it could be linked to a survival response. It seems that after domestication, fishes lost very rapidly, in only one generation, the stress response. This change in behaviour is probably directly linked to physiological changes: in the rainbow trout, two different lineages were selected on the basis of their rate of cortisol as responsiveness to stress. Individuals, which showed a low rate of cortisol, had a lower response to stress; they developed a better foraging behaviour but had a bad response to a potential danger. These individuals were well adapted to the environmental conditions of fish farms, but not the natural environment [85]. This is a general problem; the selection by humans of particular lineages of fishes based on their potentiality of growth and development has an influence on other life traits especially on behavioural traits. In the sea bass, the repetitive application of stress elements (pursuit of the fishes with a net, luminous changes, application of predator lure) modifies the foraging habits of wild fishes but also of reared ones. This could be interpreted as a habituation to the situation, which becomes less stressful [86].

8. Learning-cognition

If the domestication process leads to a change in behavioural traits, empirical evidence for a difference in cognitive performance, however, is scarce. In the framework of animal personalities, differences in behaviour may arise during ontogeny through learning and bolder, and more aggressive animals (usually, the wild form) should learn faster. Such examples exist in vertebrates especially in mammals; by comparing wild cavies and domestic guinea pigs (*Cavia porcellus)* in behavioural tests. Domestic guinea pigs were less bold and aggressive than their wild congeners, but learnt an association faster [87]. Such studies exist also in fish but are scare, and now, results are not clearly established, leading an important field of research. For example, Klefoth et al. [88] tested two common genotypes of common carp, *Cyprinus carpio* L., differing in degree of domestication (a highly domesticated mirror carp and a less domesticated scaled carp) exposed to fishing. Domesticated mirror carp were more vulnerable to angling gear than scaled carp in both environments; these results were related to a bolder-foraging behaviour for the latter. Independently of genotype, fish become more difficult to catch, indicating learned hook avoidance, based on the boldness, so scaled carp get an advantage with a lower vulnerability to fishing. The study of Rodewald et al. [89] showed that after their release in natural environment, hatchery-reared salmon had a lower foraging rate than wild individuals. They showed that this difference was the consequence of higher abilities of learning the new environment and especially the presence of potential prey by the wild fish. Such studies should be initiated before the reintroduction of hatchery stock in the natural habitat, to ensure the success of the operation.

9. Conclusion

'Domestication is that process by which a population of animals becomes adapted to man and to captive environment by genetic changes occurring over generation and environmentally-induced developmental events recurring in each generation [90]'. It affects all functions of the organisms and, in particular, behaviour. There are behavioural differences between wild and reared fishes (see **Table 1**), but these differences are more quantitative than qualitative; no new behaviours appear with domestication.

The selection of individuals for economic reasons leads to the selection of fishes on morphological or developmental traits (growth, size). These traits are directly linked to other biological traits (i.e. behaviour) and their selection may lead to select fishes, which present some behaviours affecting the life in groups of high density and so the development of each individual (increase in aggressive and agonistic interactions between individuals, higher levels of cannibalism). One solution to prevent that is to identify as soon as possible in the fish development the behavioural profiles of the individuals under different domestication levels [91]. These studies lead to better knowledge of the fish larvae, which are difficult to test given their high sensitivity to environmental conditions.

In this review, we saw that all behavioural traits may be impacted by domestication even after only one generation. For some traits, the results are clear and follow the same trend; the response to a predator is affected by domestication whatever the domesticated species and the reared environment. But in some cases, it is more difficult to find a common trend: foraging is affected but it depends on the type of food, and on the feeding conditions. It is the same for aggressiveness in the hatchery-reared individuals; it could decrease in that way we can put a high number of predators together if we give them a sufficient amount of food, but on the other hand, the high fish density in tank can produce a high level of aggressiveness between individuals leading to cannibalism event if the food is abundant. It is also true for other behavioural traits such as personality or cognitive capabilities; until now, there is a lack of studies on the influence of domestication on these behavioural traits and it is not possible to conclude. What we know is that the human selection on morphological or physiological traits of some individuals (even through a genetic program) has a direct influence on behavioural traits.

This has two implications: first, it is necessary to study behavioural traits in the case of domestication of new species in order to determine the best environmental conditions of rearing, and second, these behavioural trait modifications must be into account when release of domesticated animals into natural habitats is considered. For these two points, we have to keep in mind that the consequences of behavioural selection traits through domestication correspond to the selection of a particular behavioural trait belonging to the natural behavioural range of the species under rearing environmental conditions; this might lead to a new species, the other behavioural traits of the species range disappearing. It is known under a genetic-environment process by which the epigenetic landscape is modified by the environment constraints influencing directly the genetic program [92, 93].

Behavioural traits	Type of responses to environmental constraints	Effects of domestication	References
Swimming behaviour	Response to a predator attack	Wild is more reactive than hatchery	[10]
	Response to a novel environment	Wild exhibits higher swimming abilities	[11, 12, 13, 14]
	Capacity to be released in the wild	Wild exhibits lower dispersal	[16, 17, 18]
Foraging behaviour	Changing the foraging strategy	Hatchery exploits all the water column	[19, 20]
		Hatchery has a higher capture rate	[21, 22]
	Prey profitability	Wild has better rate than hatchery	[26]
	Consequences on survival	Wild shows a higher rate of survival	[29]
Predator avoidance	Reaction in front of a predator	Wild is more rapid	[41, 42]
	Reaction to a potential predator	Hatchery takes more risks	[36, 45]
	Spontaneous escape responses	No difference between wild and hatchery	[47]
	Use of shelter	Wild has a higher rate	[47, 48]
Social behaviour	Aggressiveness	Hatchery is more aggressive than wild	[56, 57]
		It is higher for hatchery	[55, 58]
	Agonistic behaviour	It is higher for hatchery	[58, 63, 64]
		It is higher for wild	[56, 65]
		There is no difference	[57]
	Competitive behaviour	Territoriality is higher in wild	[70, 72]
		Dominance is higher in wild	[73]
	Cannibalism	No information?	
Reproduction	Abilities to reproduce after realising	Paternity rate higher for wild	[79]
	Choice of place to spawn		
		Better for wild	[76]
		No difference	[77]
	Fitness in natural sites	Higher for wild	[71]
Personality	Boldness	Higher for hatchery	[72, 80]
		Wild higher than for wild	[82]
	QTL mapping	There is no difference in prey acquisition	[54]
	Stress	Wild is more stressful than hatchery	[83, 84]

Behavioural traits	Type of responses to environmental constraints	Effects of domestication	References
Cognitive abilities	Learning	Wild is less bolder-foraging than hatchery	[88]
	Releasing	Wild learns new environment faster	[89]

Table 1. Summary of the different results found for this review. The behavioural traits were divided in behavioural acts and their responses to domestication process in fish were briefly described. Wild referred to animals coming from wild strains and hatchery for fish larvae reared under farm or laboratory conditions.

Author details

Alain Pasquet

Address all correspondence to: alain.pasquet@univ-lorraine.fr

Domestication in Inland Aquaculture, UR AFPA—INRA, University of Lorraine, Vandoeuvre les Nancy, France

References

[1] Theodorou K, Couvet D. Familial versus mass selection in small populations. Genetics Selection Evolution. 2003;**35**:425

[2] Hamblin MT, Buckler E, Jannink JL. Populations genetics of genomics-based crop improvement methods. Trends in Genetics. 2011;**27**(3):98-106

[3] Begout Anras ML, Lagardère JP. Domestication et comportement chez les poissons téléostéens. INRA Productions Animales. Vol. 17. 2004. pp. 211-215

[4] Huntingford FA. Implications of domestication and rearing conditions for the behaviour of cultivated fishes. Journal of Fish Biology. 2004;**65**:122-142. DOI: 10.1111/J.1095-8649.2004.00562.x

[5] Mayr E. Animal Species and Evolution. Cambridge, MA: Havard University Press; 1963

[6] Kohane MJ, Parsons PA. Domestication. Evolutionary change under stress? Evolutionary Biology. 1988;**23**:31-48

[7] Jensen P. The Ethology of Domestic Animals: An Introductory Text. 3th ed. UK: CABI; 2017

[8] Huntingford FA. In: Huntingford F, Jobling M, Kadri S, editors. Aquaculture and Behaviour. Oxford, UK: Wiley-Blackwell; 2012

[9] Koolhass JM, Korte SM, DeBoer SF, Van der Vegt BJ, Van Reenen CG, Hopster H, De Jong IC, Ruis MAW, Blokhuis HJ. Coping styles in animals: Current status in behavior and stress-physiology. Neuroscience & Biobehavioral Reviews. 1999;**23**:925-935

[10] Benhaïm D, Samuel Péan S, Lucasa G, Blanc N, Chatain B. Early life behavioural differences in wild caught and domesticated sea bass (*Dicentrarchus labrax*). Applied Animal Behaviour Science. 2012;**141**:79-90

[11] Faucher K, Fichet D, Maramand P, Lagardère JP. Impact of cute cadmium exposure on the trunk lateral line neuromasts and consequences on the C-start response beahviour of the sea bass (*Dicentrarchus labrax* L., Teleostei, Monoridae). Aquatic Toxicology. 2006;**76**:278-294

[12] Johnsson N, Petersson E, Jonsson B, Bjornsson B, Jarvi T. Domestication and growth hormone alter antipredator behavior and growth patterns in juvenile brown trout *Salmo trutta*. Canadian Journal of Fisheries and Aquatic Sciences. 1996;**53**:1546-1554

[13] Lefrançois C, Domenici P. Locomotor kinematics and behaviour in the escape response of European sea bass *Dicentrarchus labrax* L. exposed to hypoxia. Marine Biology. 2006; **149**:969-977

[14] Domenici P, Batty RS. Escape behaviour of solitary herring (*Clupea harengus*) and comparisons with schooling individuals. Marine Biology. 1997;**128**:29-38

[15] Wegner NC, Drawbridge MA, Hyde JR. Reduced swimming and metabolic fitness of aquaculture-reared California yellowtail (*Seriola dorsalis*) in comparison to wild-caught conspecifics. Aquaculture. 2018;**486**:51-56

[16] Horka P, Horky P, Randak T, Turek J, Rilkova K, Slavik O. Radio-telemetry shows differences in the behaviour of wild and hatchery-reared European grayling *Thymallus thymallus* in response to environmental variables. Journal of Fish Biology. 2015;**86**:544-557

[17] Taylor MD, Laffan SW, Fairfax AV, Payne NL. Finding their way in the world: Using acoustic telemetry to evaluate relative movement patterns of hatchery-reared fish in the period following release. Fisheries Research. 2017;**186**:538-543

[18] Thompson BC, Porak WF. Using radiotelemetry to compare the initial behaviour and mortality of hatchery-reared and wild juvenile Florida bass. Transactions of the American Fisheries Society. 2016;**145**:374-385

[19] Reinhardt UG, Yamamoto T, Nakano S. Effects of body size and predators on intracohort competition in wild and domesticated juvenile salmon in a stream. Ecological Research. 2001;**16**:327-334

[20] Robinson BD, Rowland W. A potential model system for studying the genetics of domestication : Behavioral variation among wild and domesticated strains of zebra fish (*Danio rerio*). Canadian Journal of Fisheries and Aquatic Sciences. 2005;**62**:2046-2054

[21] Thodesen J, Grisdale-Helland B, Helland SJ, Gjerde B. Feed intake, growth and feed utilisation of offspring from wild and selected Atlantic salmon (*Salmo Salar*). Aquaculture. 1999;**180**:237-240

[22] Sanchez MP, Chevassus B, Labbé L, Quillet E, Mambrini M. Selection for growth of the brown trout (*Salmon trutta*) affects feed intake but not feed efficiency. Aquatic Living Resources. 2001;**14**:41-48

[23] Petersson E, Jarvi T. Both contest and scramble competitions affect the growth performance of brown trout, *Salmo trutta,* parr of wild and sea-ranched origins. Environmental Biology of Fishes. 2000;**59**:211-218

[24] Brown C, Davidson T, Laland KN. Environmental enrichment and prior experience of live prey improve foraging behaviour in hatchery-reared Atlantic salmon. Journal of Fish Biology. 2003;**63**:187-196

[25] Munaka A, Bjornsson BT, Jonsson E, Amano M, Ikuta K, Kitamura S, Kurokawas T, Aida K. Post-release adaptation processes of hatchery-reared honmasu salmon parr. Journal of Fish Biology. 2000;**56**:163-172

[26] Reiriz L, Nicieza AG, Brana F. Prey selection by experienced and naive juvenile Atlantic salmon. Journal of Fish Biology. 1998;**53**:100-114

[27] Sundström LF, Johnsson JI. Experience and social environment influence the ability of young brown trout to forage on live novel prey. Animal Behaviour. 2001;**61**:249-255

[28] Brown C, Laland N. Social learning and life skills training for hatchery-reared fish. Journal of Fish Biology. 2001;**59**:471-493. DOI: 10.1006:jfbi.2001;1689

[29] Fritts AL, Scott JL, Pearsons TN. The effects of domestication on the relative vulnerability of hatchery and wild origin spring Chinook salmon (*Oncorhynchus tshawytscha*) to predation. Canadian Journal of Fisheries and Aquatic Sciences. 2007;**64**(5):813-818

[30] Kallio-Nyberg I, Koljonen ML. The genetic consequences of hatchery rearing on life-history traits of the Atlantic salmon (*Salmo salar* L.): A comparative analysis of sea-ranched salmon with wild and reared parents. Aquaculture. 1997;**153**:207-224

[31] Johnsson JL, Abrahams LV. Inbreeding with a domestic strain increases foraging under threat of predation in juvenile steelhead trout (*Oncorhynchus mykiss*): An experimental study. Canadian Journal of Fisheries and Aquatic Sciences. 1991;**48**:243-247

[32] Berijikian BA. The effect of hatchery and wild ancestry and experience on the relative ability of steelhaed trout (*Oncorhynchus mykiss*) to avoid a benthis predator. Canadian Journal of Fisheries and Aquatic Sciences. 1995;**52**:2476-2482

[33] Dellefors C, Jonhsson JI. Foraging under risk of predation in wild and hatchery-reared juvenile sea trout (*Salmo trutta* L.). Nordic Journal of Fresh Water Research. 1995;**70**:31-37

[34] Johnsson N, Hojesjo J, Fleming IA. Behavioural and heart rate responses to predation risk in wild and domesticateAtlantic salmon. Canadian Journal of Fisheries and Aquatic Sciences. 2001;**68**:788-794

[35] Einum S, Fleming IA. Genetic divergence and interactions in the wild among native, farmed and hybrid Atlantic salmon. Journal of Fish Biology. 1997;**50**:634-651

[36] Ferno A, Jarvi T. Domestication genetically alters the antipredator behaviour of an anadromous brown trout (*Salmo trutta*) a dummy predator experiment. Nordic Journal of Freshwater Research. 1998;**74**:95-100

[37] Fleming IA, Einum S. Experimental tests of genetic divergence of farmed from wild Atlantic salmon due to domestication. ICES Journal of Marine Science. 1997;**54**:1051-1063

[38] Godin JGJ. Evading predators. In: Godin JGJ, editor. Behavioural Ecology of Teleost Fishes. New York: Oxford University Press; 1997. pp. 191-236

[39] Smith RJF. Avoiding deterring predators. In: Godin JGJ, editor. Behavioural Ecology of Teleost Fishes. New York: Oxford Universit Press; 1997. pp. 163-190

[40] Alvary D, Nieriza AG. Predator avoidance behaviour in wild and hatchery-reared brown trout: The role of experience and domestication. Journal of Fish Biology. 2003;**63**:1565-1577

[41] Nodtvedt M, Ferno A, Gjosaeter J, Stungrund P. Antipredator behaviour of hatchery-reared and wild juvenile Atlantic cod (*Gadus morua* L.) and the effect of predator training. In: Howell B, Moksness E, Svasand T, editors. Stock Enhancement and Sea Ranching. Oxford: Fishing, News, Books; 1999. pp. 350-362

[42] Berijikian BA, Tezak EP, LaRae AL. Innate and enhanced predator recognition in hatchery-reared Chinook salmon. Environmental Biology of Fishes. 2003;**67**:241-251

[43] Griffiths SW, Amstrong JD. Rearing conditions influence refuge use among over-wintering Atlantic salmon juveniles. Journal of Fish Biology. 2002;**60**:363-369. DOI: 10.1006/jfbi.2001.1846

[44] Matsuzaki SS, Mabuchi K, Takamura N, Nishida M, Whashinati I. Behavioural and morphological differences between feral and domesticated strains of common carp *Cyprinus carpio*. Journal of Fish Biology. 2009;**75**:1206-1220

[45] Sundström LF, Peterson E, Hojesjo E, Johnsson JI, Jarvi T. Hatchery selection promotes boldness in newly-hatched brown trout (*Salmo trutta*): Implications for dominance. Behavioral Ecology. 2004;**15**:192-198

[46] Yamamoto T, Reinhardt UG. Dominance and predator avoidance in domesticated and wild masu salmon *Oncorhynchus masou*. Fisheries Science. 2003;**69**:88-94

[47] Salvanes AGV. Antipredator behaviours of hatchery *Salmo salar* juveniles similar to wild juveniles. Journal of Fish Biology. 2017;**98**:1785-1796. DOI: 10.1111/jfb.13268

[48] Tang L, Jacquin L, Lek S, Liu H, Li Z, Liu J, Zhang T. Differences in anti-predator behavior and survival rate between hatchery-reared and wild grass carp (*Ctenopharyngodon idellus*). International Journal of Limnology. 2017;**53**:361-367. DOI: 10.1051/limn/2017020. (In press)

[49] Christensen KA, Brunelli JP, Wheeler PA, Thorgaard GH. Antipredator behavior QTL: Differences in rainbow trout clonal lines derived from wild and hatchery populations. Behavior Genetics. 2014;**44**:535-546

[50] Sundström LF, Lhomus M, Delvin RH. Gene–environment interactions influence feeding and anti-predator behavior in wild and transgenic coho salmon. Ecological Applications. 2016;**26**(1):67-76

[51] Pitcher TJ. Heuristic definitions of fish schoaling behavior. Animal Behaviour. 1983;**31**:611-612

[52] Shaw E. Schooling fishes. American Scientist. 1978;**66**:166-175

[53] Wood-Gush DGM, Westergaard K. Exploratory behavior and the welfare of intensively kept animals. Journal of Agricultural Ethics. 1989;**2**:161-169

[54] Wright D, Nakamichi R, Krause J, Butlin RK. Analysis of behavioral and morphological differentiation between wild and laboratory zebrafish (*Danio rerio*). Behavior Genetics. 2006;**12**:1-14

[55] Lucas MD, Drew RE, Wheeler PA, Verrell PA, Thorgaard GH. Behavioral differences among rainbow trout clonal lines. Behavior Genetics. 2004;**34**:355-365

[56] Ruzzante DE, Doyle RW. Rapid behavioral changes in medaka (*Oryzias latipes*) caused by selection for competitive and non-competitive growth. Evolution. 1991;**45**:1936-1946

[57] Ruzzante DE, Doyle RW. Evolution in a social-behavior in a ressource-rich, structured environment-selection experiments with medaka (*Oryzias latipes*). Evolution. 1993; **47**:456-470

[58] Swain DP, Riddell BE. Variation in agonistic behavior between newly emerged juveniles from hatchery and wild populations of coho salmon. Canadian Journal of Fisheries and Aquatic Sciences. 1990;**47**:566-571

[59] Metcalfe NB. Intraspecific variation in competitive ability and food intake in salmonids : Consequences for energy budget and growth rates. Journal of Fish Biology. 1986;**28**:525-531

[60] Blanchet S, Loot G, Grenouillet G, Brosse S. Competitive interactions between native and exotic salmonids: A combined field and laboratory demonstration. Ecology of Freshwater Fish. 2007;**16**:133-143

[61] Mork OI, Bjerkeng B, Rye M. Aggressive interactions in pure and mixed groups of juveniles farmed hatchery-reared wild Atlantic salmon *Salmo salar* L. in relation to tank substrate. Aquaculture and Research. 1999;**30**:571-578

[62] Ruzzante DE. Domestication effects on aggressive and schooling behavior in fish. Fisheries Aquaculture. 1994;**120**:1-24

[63] Mesa MG. Variations in feeding, aggression and position choice between hatchery and wild cutthroat trout in an artificial stream. Transactions of the American Fisheries Society. 1991;**6**:723-727

[64] Moyle G. Comparative behaviour of young brook trout of domestic and wild origin. The Progressive Fish-Culturist. 1969;**31**:51-59

[65] Hedenskog M, Petersson E, Jarvi T. Agonistic behavior and growth in newly emerged brown trout (*Salmo trutta* L) of sea-ranched and wild origin. Aggressive Behaviour. 2002;**28**:145-153

[66] Huntingford FA, Turner AK. Animal Conflict. London: Chapman & Hall; 1987

[67] Magurran AE, Seghers BH. Population differences in predator recognition and attack cone avoidance in the guppy *Poecilia reticulata*. Animal Behaviour. 1990;**40**:443-452

[68] Pichter TJ, Parris JK. Functions of schoaling behaviour in teleosts. In: Pitcher TJ, editor. Behaviour of Teleost Fishes. 2nd ed. London: Chapman & Hall; 1993. pp. 363-440

[69] Lorenzen K, Beveridge MCM, Mangel M. Cultured fish: Integrative biology and management of domestication and interactions with wild fish. Biological Reviews. 2012. DOI: 10.1111/j.1469-185X.2011.00215.x

[70] Metcalfe NB, Valdimarsson SK, Morgan IJ. The relative roles of domestication, rearing environment, prior residence and body size in deciding territorial contests between hatchery and wild juveniles salmon. Journal of Applied Ecology. 2003;**40**:535-544

[71] Berejikian BA, Tezak EP, Park L, LaHood E, Schroder SL, Beall E. Male competition and breeding success in captively reared and wild coho salmon (*Oncorhynchus kisutch*). Canadian Journal of Fisheries and Aquatic Sciences. 2001;**58**:804-810. DOI: 10.1139/cjfas-58-4-804

[72] Sundström LF, Lhomus M, Johnsson JI. Investment in territorial defence depends on rearing environment in brown trout (*Salmo trutta*). Behavioral Ecology and Sociobiology. 2003;**54**:249-255

[73] Pearsons TN, Fritts AL, Scott JL. The effect of hatchery domestication on competitive dominance of juvenile spring Chinook salmon (*Oncorhynchus tshawytscha*). Canadian Journal of Fisheries and Aquatic Sciences. 2007;**64**:803-812

[74] Campbell JM, Carter PA, Wheeler PA, Thorgaard GH. Aggressive behavior, brain size and domestication in clonal rainbow trout lines. Behavior Genetics. 2015;**45**:245-254

[75] Baras E. Cannibalism in fish larvae: What have we learned? In: Qin JG, editor. Larval Fish Aquaculture. New York: Nova Science Publishers [ISEM 2012-053]; 2012. pp. 1-37

[76] Johnsson N, Johnsson B, Hansen LP. Partial segregation in the timing of migration of Atlantic salmon of different ages. Animal Behaviour. 1990;**40**:313-321

[77] McLean JE, Bentzen P, Quinn TP. Differential reproductive success of sympatric, naturally spawning hatchery and wild steelhead trout (*Oncorhynchus mykiss*) through the adult stage. Canadian Journal of Fisheries and Aquatic Sciences. 2003;**60**:433-440

[78] Unwin MJ, Glova GJ. Changes in life history parameters in a naturally spawning population of Chinook salmon (*Oncorhynchus tshawytsha*) associated with releases of hatchery-reared fish. Canadian Journal of Fisheries and Aquatic Sciences. 1997;**54**:1235-1245

[79] Neff BD, Garner SR, Fleming IA, Gross MR. Reproductive success in wild and hatchery male coho salmon. Royal Society Open Science. 2018. DOI: 10.1098/rsos.150161

[80] Sih A, Bell A, Johnson JC. Behavioral syndromes: An ecological and evolutionary overview. Trends in Ecology & Evolution. 2004;**19**:372-378. DOI: 10.1016/j.tree.2004.04.009

[81] Berejikian BA, Tezak EP, Schroder SL, Flagg TA, Knudsen CM. Competitive differences between newly emerged offspring of captive-reared and wild coho salmon. Transactions of the American Fisheries Society. 1999;**128**:832-839

[82] Brignon WR, Pike MM, Ebbesson LOE, Scholler HA, Peterson JT, Schrek CB. Rearing environment influences boldness and prey acquisition behaviour, and brain and lens development of bull trout. Environmental Biology of Fishes. 2018;**101**:383-401

[83] Price EO. Animal Domesticaion and Behavior. New York: CABI Publishing; 2002

[84] Walker JA, Ghalambor CK, Griset OL, Kenney DM, Reznick DN. Do faster starts increase the probability of evading predators? Functional Ecology. 2005;**19**:808-815

[85] Overli O, Pottinger TG, Carrick TR, Overli E, Wenberg S. Differences in behaviour between rainbow trout selected for high and low-stress responsiveness. Journal of Experimental Biology. 2002;**205**:391-395

[86] Millot S, Pean S, Leguay D, Vergnet A, Chatain B, Begout ML. Evaluation of behavioral changes induced by a first step of domestication or selection for growth in the European sea bass (*Dicentrarchus labrax*): A self-feeding approach under repeated acute stress. Aquaculture. 2010;**306**:211-217

[87] Brust V, Guenther A. Domestication effects on behavioural traits and learning performance: Comparing wild cavies to Guinea pigs. Animal Cognition. 2015;**18**(1):99-109

[88] Klefoth T, Pieterek T, Arlinghaus R. Impacts of domestication on angling vulnerability of common carp, *Cyprinus carpio*: The role of learning, foraging behaviour and food preferences. Fisheries Management and Ecology. 2013;**20**:174-186. DOI: 10.1111/j.1365-2400.2012.00865.x89

[89] Rodewald P, Hyvarinen P, Hirvonen H. Wild origin and enrichment environment promote foraging rate and learning to forage on natural prey of captive reared Atlantic salmon parr. Ecology of Freshwater Fish. 2011;**20**:569-579

[90] Price EO. Behavioral development in animals undergoing domestication. Applied Animal Behaviour Science. 1999;**65**:245-271

[91] Teletchea F, Fontaine P. Levels of domestication in fish: Implications for the sustainable future of aquaculture. Fish and Fisheries. 2014;**15**(2):181-195. DOI: 10.1111/faf.12006

[92] Waddington CH. The Strategy of the Genes; a Discussion of Some Aspects of Theoretical Biology. London: Allen & Unwin; 1957

[93] Golberg AD, Allis CD, Bernstein E. Epigenetics: A landscape takes scape. Cell. 2007; **128**(23):635-638

Chapter 6

Domestication and Welfare in Farmed Fish

João L. Saraiva, Maria F. Castanheira,
Pablo Arechavala-López, Jenny Volstorf and
Billo Heinzpeter Studer

Additional information is available at the end of the chapter

http://dx.doi.org/10.5772/intechopen.77251

Abstract

The domestication of fish species is still in its early stages when compared to terrestrial animals. The effects of domestication on welfare of farmed fishes are complex to study because fish differ from livestock in genetics, physiology and behaviour, and experience different sensory worlds. Consequently, empathy with fish and understanding of their needs becomes more problematic than with land animals. Additionally, the acknowledgement and study of mental dimensions of fish existence is very recent. We discuss that higher levels of domestication in fish do not necessarily correspond to better welfare because (1) artificial selection by the aquaculture industry is mostly focused on production-related traits such as growth, and this selection process may have unknown negative effects on welfare-related traits; (2) the number of fish species presently farmed (*circa* 300) is 10-fold higher than land animals, rendering the establishment of standard welfare guidelines extremely complicated; (3) the current paradigm of the Five Freedoms guiding welfare is out-dated and was designed for livestock; and (4) there are still severe knowledge gaps in the biology of farmed fishes, especially in welfare-related traits. The implementation of humane farming systems should integrate industry, science and ethics in an open dialogue in order to produce relevant results.

Keywords: artificial selection, fish, behaviour, aquaculture, welfare

1. Introduction

1.1. Domestication

Domestication is a human-induced process that gradually changes a cultured organism. It extends over generations and involves developmental effects within each generation, culminating generally in genetic changes across generations [1]. One of the first domesticated animals was the goat (*Capra hircus*), in a process that started over 10,000 years ago in Iran [2]. Present-day domesticated goats are therefore the product of a long-lasting human selection of animals and breeders carrying specific traits in meat, fur, milk and docile behaviours, and also natural selection for fitness under variable conditions [3]. To put things in perspective, fish domestication is a much more recent process: while exceptions such as carp (*Cyprinus carpio*), tilapia (*Oreochromis niloticus*) and goldfish (*Carassius auratus*) may have possibly been artificially selected for hundreds of years, and another 12 species such as cod (*Gadus morhua*), salmon (*Salmo salar*) and trout (*Oncorhynchus mykiss*) in the 1800s [4], the vast majority of farmed fish has been under domestication merely since the middle of the twentieth century [5–8].

Although the usual interpretation of animal domestication has been related to genetic changes induced by artificial selection [9], developmental effects have been gathering more interest recently due to the increasing number of domesticated species and purposes of captive breeding [1]. These developmental responses are especially important in fishes because they present remarkably high levels of phenotypic plasticity, much more than terrestrial vertebrates [10, 11]. In Sections 2.1 and 3.1 of the current text, we will dive deeper into these matters. But in fact, this plasticity allows fish to adapt their phenotype to rearing conditions, which, taking into account their recent domestication process, makes fish domestication a controversial issue: although there is variation in morphology, physiology and genetics between farmed fish species and their wild conspecifics [12–14], it is generally much lower than between terrestrial farm animals and their wild counterparts [4, 9, 11, 12]. Nevertheless, it is mostly established that fish farmed under well-managed systems (i.e. that provide conditions that enhance growth and survival while supplying the correct nutrition regimes for the species) can maximise growth to nearly their physiological maximum, suffer lower mortality rates than in the wild and are usually less prone to infectious diseases [10].

The domestication process tends to produce a coherent set of behavioural, morphological and physiological changes that are similar in many species. A typical domesticated phenotype of a species could therefore be summarised as differing from its wild ancestor in having a different colour (probably being brighter or spotted); being brachycephalic, chondrodystrophic and smaller; having a reduced brain size, earlier sexual maturation and increased reproduction; with a faster and more flexible development; and being less fearful, more sociable, and more risk-prone towards predators. This is a recurrent trait complex that suggests that it may represent a general adaptation pattern to captivity and domestication [1, 15]. Throughout this chapter, we will review these changes and if and how they can affect the welfare of farmed fish.

The domestication of farmed fish is therefore not straightforward, as it depends on the definition of domestication, and how literally that definition is applied to each case. To tackle this difficulty, some authors proposed that domestication in fish should be considered as a process, with progressive stages summarised in **Table 1** [8].

Throughout this chapter and for streamlining purposes, we will adopt the nomenclature given by the authors of the papers cited (e.g. domesticated, artificially selected, strain, etc.), regardless of the stage they appear to be in the classification proposed in **Table 1**.

http://dx.doi.org/10.5772/intechopen.77251

Domestication level	Description	Examples from FishEthoBase (see Section 4 for a description)
1	Acclimatisation to the culture environment	Malabar trevally (*Carangoides malabaricus*)* Yellowback sea bream (*Dentex tumifrons*)* Spotted sea bass (*Dicentrarchus punctatus*)*
2	Part of the life cycle is completed in captivity, but several important bottlenecks still exist in others (e.g. reproduction, larval rearing, etc.)	Wreckfish (*Polyprion americanus*) Yellowtail amberjack (*Seriola lalandi*) Greater amberjack (*Seriola dumerili*)
3	The entire life cycle is closed in captivity, but with wild inputs	Pangasius (*Pangasianodon hypophthalmus*) Southern bluefin tuna (*Thunnus maccoyii*) Burbot (*Lota lota*) Atlantic halibut (*Hipoglossus hipoglosus*) Senegalese sole (*Solea senegaensis*) Turbot (*Scophthalmus maximus*)
4	The entire life cycle is closed in captivity without wild inputs, but no selective breeding programme is used	Atlantic cod (*Gadus morhua*) Cobia (*Rachycentron canadum*) Meagre (*Argyrosomus regius*) African catfish (*Clarias gariepinus*) Russian sturgeon (*Acipenser guelden-staedtii*) Adriatic sturgeon (*Acipenser naccarii*)
5	Selective breeding programme is used focusing on specific goals (growth rate, fillet yield, flesh quality, etc.)	Common carp (*Cyprinus carpio*) Nile Tilapia (*Oreochromis niloticus*) Gilthead seabream (*Sparus aurata*) European seabass (*Dicentrachus labrax*) Atlantic salmon (*Salmo salar*) Rainbow trout (*Oncorhynchus mykiss*) Siberian sturgeon (*Acipenser baerii*)

*Soon in FishEthoBase.

Table 1. Levels of fish domestication. Adapted from [8].

1.2. Welfare

Definitions of welfare generally vary between the function-based and the feelings-based approaches [16]. Function-based approaches look more into the biological, physiological and overall health perspective of the animal [17]. Feelings-based approaches, on the other hand, link welfare with the emotional (or emotional-like) state of the animal [18] and define welfare in a longer term as the balance between positive and negative subjective experiences [19]. While the first approach might may be too narrow (e.g. a man in prison may be in perfect health and still be experiencing terrible welfare) and the second too wide (how to measure it?), a behavioural approach may solve the question of incorporating both physiological and mental indicators. Even more, it operationalizes the concept and allows objective measures of welfare [16]. In that sense, we may define welfare as the state of the individual as it copes with the environment [20]. This definition of welfare has several implications: (i) welfare is a characteristic of an animal, not something that is given to it; (ii) welfare will vary from very bad to very good, i.e. along a welfare continuum; (iii) welfare can be measured independently of ethical considerations; (iv) measures of failure and difficulty to cope with the environment give information about how poor the welfare is; (v) knowledge on the biology and life-history of an animal provide essential information about suitable rearing conditions, but direct measurements of the state of the animal must also be used to assess its welfare; and (vi) coping mechanisms may vary among different species, and there are several consequences of failure to cope. Therefore, any one of a variety of measures can indicate that welfare is bad, and the fact that one measure, such as growth, is normal does not mean that welfare is good [21].

2. Natural and artificial selection

In order to understand the effects of artificial selection on welfare-related traits, we must first address natural selection. Although selection acts on differences in survival and reproductive success between individual organisms, or phenotypes, what changes during evolution is the relative frequency of genes. Therefore, Darwin's theory in modern terms may be stated as follows:

1. All organisms have genes coding for proteins and regulating the development of everything in the organism, including the nervous system, muscles and structure of the individual—and so influence its behaviour.
2. Within a population, most genes are present in two or more forms, or alleles, which code for slightly different forms of the same protein or determine when, where and how much of the protein is expressed. These will cause differences in development and function, and so there will be variation within a population.
3. Any allele that results in more surviving copies of itself than its alternative will eventually replace the alternative form in the population. Natural selection is the differential survival of alternative alleles through their effects on replication success.

The individual can be regarded as a temporary vehicle or survival machine by which genes survive and replicate [22]. Because selection of genes is mediated through phenotypes, the most successful genes will usually be those that are most effective in enhancing an individual's survival and reproductive success [23].

The actual basis of selection processes is variation, which is widespread in all biological phenomena. Variations within species are called polymorphisms, and these are (at least partly) independent of ontogeny and sex. This variation is genetically based and heritable [24]. Conversely, phenotypic plasticity can be broadly defined as (i) the ability of one genotype to produce more than one phenotype when exposed to different environments, (ii) the modification of developmental events by the environment or (iii) the ability of an individual organism to alter its phenotype in response to changes in environmental conditions [25]. Domesticated strains (lineages or populations that have been under artificial selection for generations and that, at least empirically, seem to differ from its wild origin) rely on natural polymorphisms to be selected: for example, present-day domesticated goats are the product of 10,000 years of artificial selection on breeders for specific traits in meat, milk and fur, as well as of natural selection for fitness under variable conditions [3]; fast growing Atlantic salmon (*Salmo salar*) is the result of 40 years of research and artificial selection for fast growing progeny, with an increase of 10–15% in each generation [26].

Although the gap between the time domestication has been underway in terrestrial farm animals and in farmed fish is enormous, there has been a considerable effort towards fish domestication in recent times [4]. However, the main component in the domestication process is the generation interval (i.e. the average age of the parent animals at the birth of their offspring—note that this is *not* the age at maturity). A comparison between land animals and fish may be found in **Figure 1**.

While pig (*Sus scrofa domesticus*), sheep (*Ovis aries*), goat and horse (*Equus ferus caballus*) vary between 1.5 and 3.5 years [27], in fish we may find generation intervals from 6 to 8 months in tilapia [28], 3–4 years in salmon and trout [26], 4–6 years in sea bass and sea bream [29, 30] and eventually from 12 to 33 years in some species of sturgeon [31]. This means that generation intervals may be in the same order of magnitude in terrestrial animals and fish, but also shows how variable these can be for different species in aquaculture. The variation in the average age at reproduction poses a challenge for the establishment of general standards of welfare in fish farming: for example, designing a welfare plan for a tilapia farm will differ immensely from a sturgeon facility.

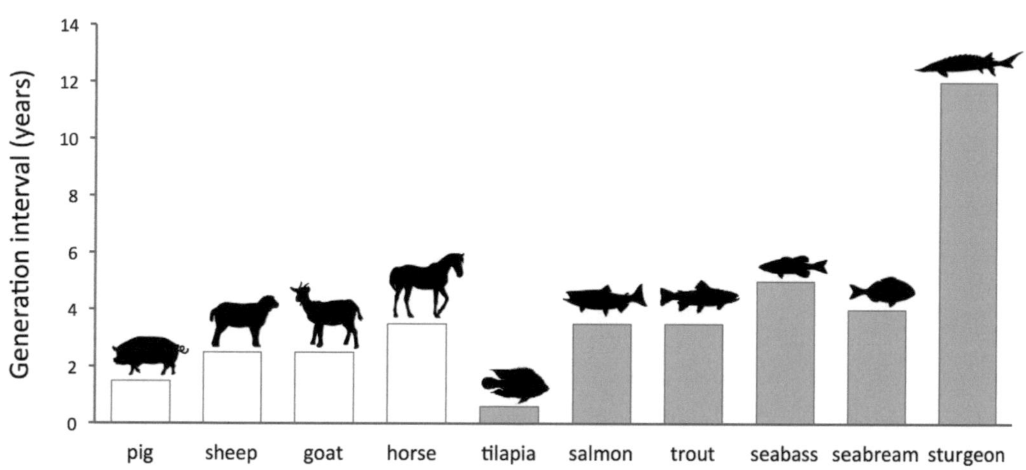

Figure 1. Comparison between the generation intervals in terrestrial farm animals (white bars) and fishes (grey bars).

3. Domestication and welfare

3.1. Genes

In the artificial conditions provided by human farming activities, it is likely that the most successful phenotypes of farmed animals are different from those under natural conditions. However, that does not mean that these selected phenotypes carry differential genotypes, especially due to phenotypic plasticity which is remarkably relevant in fish: different populations of the same species present contrasting yet plastic behavioural responses to environmental and social conditions [32–38]. Nevertheless, artificial selection experiments demonstrate that almost any quantitative trait could be permanently altered, that responses (mostly) occurred as a consequence of changes in the frequencies of genes affecting the traits, and not from mutations, and that many genes must be involved [39, 40].

In livestock species, genetic selection has greatly increased production levels. Usually, the breeding goal is to create a population with high economic production efficiency, i.e. high production combined with relatively low feed intake. Breeding programs have become quite successful because of the high accuracy of breeding value estimation, the moderate to high heritabilities of most production traits and the use of large and fast databases containing production records of many animals and their genetic relationships. Apart from genetic changes, production is also increased by improvement of housing, feed composition, feeding strategies, health status and farm management.

However, negative side-effects of domestication largely occur and are expected to increase when the focus continues to be only on production efficiency. Animals in a population that has been genetically selected for high production efficiency seem to be more at risk for behavioural, physiological, immunological, reproductive and consequently welfare problems [41]. This occurs because behavioural traits, as well as the other typical components of the domestication phenotype (growth, stress, immune function, etc.), are most likely controlled by many genes, i.e. they are polygenic [15]. In addition, a given set of genes may influence different traits, a mechanism known as pleiotropy [42]. In such a case, increasing the frequency of alleles that, for example, up-regulate growth, may at the same time modify other essential welfare-related traits under the influence of the same genes. Finally, the function of one gene may also be influenced by the interaction with other genes, which is known as epistasis [43]. Selection for one or a few traits controlled by genes that have epistatic effects may thus influence a group of other genes, regulating other characters than those selected for. Therefore, both mechanisms (pleiotropy and epistasis) are more than likely to create side-effects on traits that are not desirable [15]. In fact, when animals are selected for production traits mainly, many side-effects have been extensively documented in several species [41]. Some of these side-effects will even affect production itself, such as reduced fertility in fast- growing broilers [41], and are likely to be the target of counter-selection. In other cases, the side-effects may be related to less-obvious traits which may nevertheless have a strong welfare aspect [44] because they shift welfare optima towards unknown directions. These processes have occurred in land animals throughout their domestication. However, the slow and long domestication process on land has allowed both humans and many livestock species to adapt and cope with such effects. Conversely, the low number of generations in the rapidly increasing, diverse and generally very recent fish farming activity may be too short to permit this adaption in aquatic species.

3.2. Physiology

Changes in the phenotypes of selected farmed fish usually correlate with changes in physiological indicators. For example, when comparing seventh-generation farmed Atlantic salmon with wild individuals, the domesticated fish grow much faster (even more so in salt water where the difference is threefold), pituitary and plasma growth hormone levels were positively correlated with growth rate and significantly higher in the domesticated strain [45]. The same occurs with strains of Brook trout (*Salvelinus fontinalis*) farmed for 30 or more years (ca. 10 generations). The domesticated phenotype is less resilient than the wild, since hatchery-born fingerlings struggle to survive when released into native streams. Moreover, domesticated strains grow better in aquaculture but not in the wild [46], which led some authors to claim that domesticated animals are better adapted to captivity, reducing stress and mortality, increase disease resistance, reduce the use of chemotherapeutics and contribute to better animal welfare and environmental management. Up to 65% of farmed salmon in Norway comes from improved breeding plans, after an extensive programme lasting over 40 years and an enormous investment effort, with numbers reaching nine digits [26]. However, and despite these efforts and claims, 40 years represent only 10–13 generations of farmed salmon (**Figure 1**). Even more importantly, a recent analysis of welfare conditions of farmed salmon revealed a need for improvement in space, substrate, aggression, stress levels and malformations ([47], see *Salmo salar*). A possible conclusion is that the improvement programme of salmon in Norway may be focusing mostly on production-related traits. Although there are improvements on the health perspective of welfare, the natural needs and behaviour of this species in captivity may be generally impaired.

For many physiological indicators such as those listed below, selected strains of farmed fish fail to show positive results:

Metabolic rate: fast growing hatchery strains of Rainbow trout present higher standard metabolic rate (SMR), lower aerobic scope, and potentially lower maximum metabolic rates, suggesting that high growth trades off against a reduced capacity to do metabolic work. Higher SMR of fast growers appears to be related to a greater investment in high-maintenance digestive tissue that supports rapid growth, which appears to compromise active metabolism [48]. Farmed Senegalese sole (*Solea senegalensis*) born from wild spawners are nevertheless capable of shifting their routine metabolism from naturally nocturnal to diurnal, responding to daylight feeding regimes [49].

Hypoxia resistance: triploid strains of domesticated Rainbow trout show faster growth than wild diploid individuals probably due to impaired gametogenesis of 3n fish [50]. Adding to these reproductive problems, triploids are also less resistant to hypoxia [51], which can account for lesser resistance and higher mortalities both in nature [52, 53] and in several types of farming conditions and methods [54–57].

Stress: there are reports of selected strains of farmed fish showing lower stress responses to acute artificial stressors. In Rainbow trout, the cortisol responses to confinement in a net or to electroshock are higher in wild fish than in hatchery-reared animals [58]. However, there are also examples of the opposite pattern, even in the same species: wild trout show lower physiological stress responses to hooking than domesticated trout [59]. Immediate cortisol response to an acute handling stressor in a domesticated Eurasian strain of common carp is also higher than wild Japanese strain [60]. Although cortisol is a universally used indicator for stress, it should be used with caution. This hormone has an adaptive physiological role and several factors (e.g. genetic, developmental, environmental, frequency of exposure) underlie its release upon stressful episodes [61]. An appreciation of these factors, along with knowledge of the biology of the species, solid behavioural observations [16] and secondary stress indicators, such as plasma glucose and lactate [61], are essential for proper interpretation of the data and design of mitigation measures.

Overall performance: wild and hybrid (domesticated × wild) strains of Brook trout showed better rates of recovery (from angling) and yield than a domestic strain [62]; wild strains of Nile tilapia (*Oreochromis niloticus*) perform as well as domesticated strains [63, 64]. Triploid strains of trout also generally tend to have higher malformation rates than wild populations [65].

From the available data on these indicators, it is not clear whether the current domestication process brings any obvious and effective physiological welfare benefits.

3.3. Behaviour

As occurs with terrestrial farm animals, the environment experienced by cultured fishes highly differs from the wild [14, 66, 67]: the physical environment is much simpler, space is restricted, and migration is not possible; food is readily available so long distance tracking of food is unnecessary; there are generally fewer or no predators (apart from human), and they are treated for some diseases. For parent animals, reproduction occurs without the need to compete for mates as it is often the case in the wild. In these aspects, the environment is overall less challenging. In others, however, it is more challenging: fishes are frequently disturbed by human activity, they are usually confined at unnatural densities, which potentially increases the risk of infection and the incidence of social encounters, including aggressive ones, especially when competing for food. The hatchery environment is so different from that experienced in nature that it can potentially generate behavioural differences in three, interlinked ways: (1) differential experience, (2) differential mortality and survival of behavioural phenotypes within a single generation and (3) selection for inherited behavioural traits over several generations [68]. In fact, usual conditions in intensive husbandry favour risk-taking/aggressive fish, as available data suggests that competition for food is major driver for high-risk/high-aggression phenotypes [69]. Hatchery-reared fish are more prone to show higher risk-taking behaviour, which is directly linked with a higher risk of escapes from rearing systems (as in the case of sea cages) [70], and may severely decrease their chances of survival in the wild. Consequently, this leads to a wide range of welfare, environmental and economic consequences. Domesticated strains of guppies (*Poecilia reticulata*), which are not farmed for food but are nevertheless the object of strong artificial selection for aquarium hobbyists, tend to have smaller brains and less cognitive abilities.

This may not only be due to a trade-off between brain size and reproduction [71], but also to the lack of environmental challenges and corresponding cognitive selection pressures.

Despite the consistency of the farming environment, different coping styles (e.g. consistent trait associations such as proactive, active coping or bold, and reactive, passive coping or shy) do emerge in farmed fish [72]. The proactive/reactive continuum has been identified in most farmed species and it should be a factor to consider when designing and evaluating production systems. However, the aquaculture industry selects mostly for growth performance [73] and proactive fish grow faster [74]. Consequently, there is a theoretical infinite selection for proactive and aggressive individuals in fish farming. This creates an obvious welfare problem that can only be solved by a deep understanding of the biology of the species as well as through the design of appropriate and diverse farming environments, which can accommodate different coping styles, even at the expense of lower production outputs [72]. In addition to coping styles, there is another source of intrinsic variability in animals that is phenotypic plasticity, best explained by behavioural reaction norms (BRN, i.e. the set of behavioural phenotypes that a single individual produces in a given set of environments) [75]. The BRN may actually be calculated, and it incorporates information on how an animal behaves on average and how its behaviour changes over a gradient, specifying the precise form of the relationship between response value and environmental condition. The relationships between food provisioning rate and begging intensity, between dispersal behaviour and current velocity, or between anti-predator behaviour and predation risk are all examples of BRNs. This approach treats both inter-individual and intra-individual variance in behaviour as meaningful (rather than as 'noise') [76].

The implications of the domestication process on the behavioural perspective of welfare are therefore far from simple. Behavioural changes due to generations in captivity do seem to occur but (1) they are accompanied by physiological and cognitive modifications that are challenging to accommodate in good welfare, and (2) while the behavioural phenotypes of wild fish are adaptive and selected throughout stable evolutionary pressures, captive phenotypes are responding to extremely different settings that are artificially rapid and that can often push welfare needs into collision with traits required for production.

The available evidence, however, is largely based on data from salmonids. These species are nonetheless far from representing the majority of production of finfish in global aquaculture: Atlantic salmon ranks seventh in production worldwide with approximately 2.4 million tonnes in 2015 (less than half of the production of the #1, Grass carp (*Ctenopharyngodon idella*) with 5.8 million tonnes) [77].

3.4. Sensory worlds

Fish are an extraordinary group of animals. Our 'underwater cousins', as Jonathan Balcombe describes them in his book *What a fish knows* [78], are the closest living relatives to our common aquatic ancestor. But, as with most distant family members, our understanding of their lives is limited. Even though ichthyology was incorporated as a formal science by Aristotle (383–322BC) [79], we still struggle to understand many aspects of fish biology. Those limitations to our knowledge of fish arise mainly from the fact that fish live in water. This posts a strong barrier for the direct observation of these animals, and up until recently the study of fish was restricted to investigation from the surface under particular conditions of water transparency and shallow depth, the examination of dead specimen or watching captive animals in artificial conditions. This constraint was only truly overcome with the invention of the self-contained underwater breathing apparatus by Jacques-Yves Cousteau and Emille Gagnan in 1942. Thus, only roughly 80 years ago could humans consistently observe fish in their natural habitats, in a similar way than we had been doing with terrestrial animals since the dawn of our species. This gap in the knowledge of fish biology is a major drawback for the establishment of welfare standards. Since self-experience and individual knowledge are impossible to be observed directly, their existence in other species tends to be forgotten or ignored, especially in taxa with which we do not readily identify or that are distantly related to us [80]. While we as humans can easily empathise with cattle, goats, sheep, horses and other terrestrial animals because they have been living next to us for millennia and share most of our sensory world, fishes exist in a realm of their own.

In fact, there are likely to be substantial differences in fish sensory systems compared with a terrestrial animal due to differing ecological and evolutionary pressures [81]. The term *Umwelt* was coined by Jakob von Uexküll in 1909 and refers to the sensory world of an animal—i.e. a *subject*—who is perceiving and actively responding to environmental stimuli. Moreover, the animal is not reacting mechanically to the world, and instead building its *Umwelt* with a meaningful living strategy, even though the behaviours may not be consciously planned [82]. This concept is of vital importance for the design of welfare solutions for captive fishes, because the sensory world of these animals differs highly from our own experience, is extraordinarily diverse, and relies on senses that differ from ours.

3.4.1. Vision

Light behaves differently underwater than at the surface and can be influenced by physical and biological factors. Depth can modulate the wavelength (i.e. the colour), while intensity and scatter can be modified by turbidity and suspended particle type. These can also change rapidly with daytime, season or weather conditions. Furthermore, species have different visual systems depending on their life-history (e.g. predators that rely on visual cues for feeding, fishes that are common preys and must remain vigilant for evasion) or even within life stages (e.g. larvae that live in the depths and move to shallower depth when they grow, species with ocean juveniles and freshwater adults). These environmental changes represent huge selective pressures for the radiation of visual systems in fish. Not surprisingly, there is an enormous variety not only in the type of eyes that can be found in fish [83], but also in the brain structures that process visual information [84].

3.4.2. Chemical sensing

Chemical senses serve an essential ecological role and are extremely relevant in communication contexts in all groups of fish (cyclostomes, elasmobranchs and teleosts). They enable orientation in the dark or blurry waters, predation, foraging and escape from predators for example [85]. Chemical sensing also serves intra-specific communication, allowing males and females to find suitable partners [86], as well as competitors to assess and announce their status in agonistic contexts, which are solved much quicker and less violently thanks to 'chemical diplomacy' [87, 88].

Chemical sensing in fish exists in three modalities: olfaction, solitary chemosensory cells and taste. Olfaction may function at a larger distance for all the roles described above, and olfactory receptors are usually located in the nostrils on the most anterior part of the head. Taste cells are usually limited to very close range detection of foodstuffs and are located in the head and mouth [85]. In most predatory fishes, the taste system is used solely during oral food evaluation [89]. Solitary chemosensory cells are not well described yet but may serve as food, predator or conspecific locators, spread throughout the body of the animal [85].

3.4.3. Hearing

In all fishes, sound is detected by one or more of the otolith organs. As sound passes through a fish and brings its tissues into motion, the otoliths respond to sound-induced motions of the animal's body. In many fish species, named hearing specialists, the otoliths may also receive a displacement input from the swimbladder or another gas-filled chamber near the ears. These fishes may respond to both acoustic pressure and particle motion with a particularly efficient coupling between the gas bladder and the otolith organs and tend to have very high sensitivity to sound [80].

In addition, fishes have evolved a diversity of sound-generating organs. These include vibrating the swimbladder and pectoral girdle or rubbing bony elements against each other. Sounds are produced in various behavioural contexts (agonistic interactions, courtship, spawning and in distress). Similarly to chemical communication, acoustic signals may serve in decreasing aggression, assessment of the fighting abilities, species recognition, mate attraction and mate choice [90].

3.4.4. Nociception

The aquatic environment influences basic perception and adaptation to damage in fishes: for example, they cannot fall because of buoyancy in the water column and this prevents injury due to gravity; noxious chemicals entering the aquatic environment may be diluted and thus pose a lower risk; and major shifts in temperature are less common compared with terrestrial environments. This could mean fishes experience less risk of damage than terrestrial animals, and it may be reflected in their nociceptive system [81]. In fact, although receptors for damaging stimuli have been found in all fish groups, and fishes possess neuroanatomical pathways comparable to those found in other vertebrate groups, there are interesting differences that reveal adaptions to evolutionary pressures: for example, rainbow trout nociceptors are not activated in low temperatures, because they live in cold water [91], but they are more sensitive to mechanical stimuli than mammals, probably because their skin is more fragile, and to heat, probably because they live in temperatures usually not above to 25°C [92]. The Chameleon cichlid (*Australoheros facetus*), on the other hand, is far more tolerant to heat exposure, which can also be explained by its broad ecological distribution [93]. Importantly, fish are ectothermic, and therefore their inner temperature depends on the environment (typically 0–30°C). As mammals maintain homeostasis at 37°C, it is likely that fish nociceptors have a lower temperature threshold than mammals [81].

Interestingly, the same groups of substances that reduce pain in humans (opioids, anti-inflammatory drugs and local anaesthetics) are also effective in reducing behavioural and physiological indicators of discomfort in teleosts, which is indicative of similar sensing mechanisms [81].

3.4.5. Other senses

There are sensory systems in fish that are completely alien to us. The lateral line for example, which serves as a receptor for hydrodynamic stimuli such as those generated by conspecifics, predators or prey. Although the biological processing of hydrodynamic signals has been well studied, not much is known about how fish can discern these from natural occurring events [94]. As all fishes experience night, darkness or turbid waters, there is strong selection for the use of non-visual senses in all fish species. Anatomical diversity suggests that the lateral line is one of the most important senses for fishes. However, research on the function of the lateral line has lagged due to poor understanding of hydrodynamics at small scales and lack of this sense in humans, making it difficult to imagine a fish's hydromechanical world [95]. Electrical sensing is ancestral to fishes and is present in most non-teleosts as well as certain teleost species. The electrosensory world of fishes is rich with electric fields from a multitude of sources including the earth's magnetic field and the bodies of all aquatic organisms including the electrosensing fish itself. The fish's extremely high sensitivity to these fields enables orientation, navigation, communication, and even detection and localization of other fish, both prey and conspecifics [96–98]. **Figure 2** summarises the sensory world of fish.

Not only the sensory world of fishes is difficult to relate to, but also the physics of movement underwater in a three-dimensional world can be challenging to understand for humans, who exist roughly in a 2D world. Despite this challenge, it is nonetheless a critical next step for the

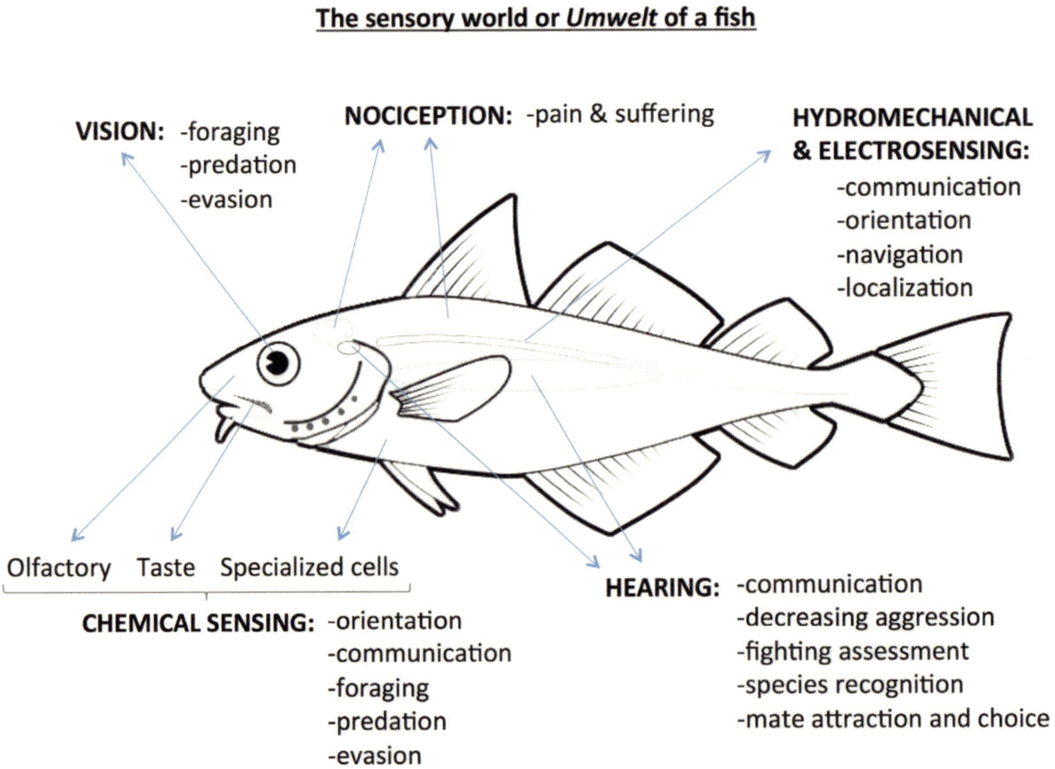

Figure 2. The sensory worlds of fish.

understanding of fish locomotion, and the design of appropriate rearing systems. The 3D nature of fish functional design is clearly demonstrated in the enormous diversity of body shapes and swimming modes in fishes [99].

Other physical properties of water affect fish in a different way than dry land does to farm animals: for example, water is a dense medium, so fish are constrained by hydrodynamic demands and fast swimming can be costly; gases dissolve readily in water, but moving water for oxygen extraction is energetically costly; many other chemicals readily dissolve and disperse in water.

In order to adapt to such a different medium, fishes not only develop the extraordinary sensory systems we have discussed above, but also show many amazing morphological and physiological adaptations, that can strongly determine welfare needs: they may undergo dramatic changes in form and function across life stages, as in the case of flatfishes [100]; unlike mammals but in common with birds, fish red blood cells are nucleated, giving them additional functions including immune responses [68]; fish grow continuously [101], influencing their relation with space and density across time; most species excrete ammonia (which is highly toxic, especially in aquaculture conditions [102]) while land animals excrete urea [103]. Finally, fish have more genes, more gene variability and more gene duplicates than terrestrial animals [104, 105].

To summarise, the extraordinary features of the aquatic environment, the exotic adaptations of fish and their *Umwelt* represent a challenge for the assessment of fish welfare. Only through a deep understanding not only of the fundamental differences between fish and terrestrial farm animals but also of the specific needs of each species can we design appropriate measures to improve and establish high standards of welfare in aquaculture. This task becomes even more daunting considering the number of animal species currently being farmed in aquatic environments: 362 finfishes (including hybrids), 104 molluscs, 62 crustaceans, 6 frogs and reptiles, and 9 aquatic invertebrates [106]. For the sake of comparison, there are 26 well-studied species of terrestrial farm animals, according to the Domestic Animal Diversity Information System (DAD-IS) [107]. Due to their low number of species, welfare measures and standards are easier to establish for land animals than for fishes. With such a long list of fish species in current world farming, how to tackle the issue of assessing fish welfare in a global manner?

4. Assessing welfare in farmed fish

Welfare in aquaculture has been a motive of academic work in the recent past. Several authors have addressed the topic in reviews and research papers [16, 68, 108–110], and the COST action *Welfare of fish in European aquaculture* has been promoted aiming to (i) improve the knowledge on welfare of fish, (ii) formulate a set of guidelines embodying a common and scientifically sound understanding of the concept of welfare in farmed fish, and (iii) construct a range of targeted operational welfare indicator protocols to be used in the industry [111]. The results of this action were incorporated in many research projects, not only in Europe but also in the USA, Canada, and New Zealand. In addition, major stakeholders in the industry were also involved, including the European Aquaculture Associations, the Fish Farmers Association and the scientific department of EU responsible for fish welfare.

In recent years, the FishEthoBase project (http://fishethobase.net) has been working in order to bridge the gap even further between the scientific community and the fish farming industry. This open-access database on fish ethology and welfare provides a platform where scientific knowledge is scrutinised and summarised in order to answer relevant criteria regarding welfare in aquatic animal farming. The aim is to cover all fishes farmed nowadays as well as other aquatic species, delivering concrete solutions for fish farmers, pointing to knowledge gaps for researchers and providing awareness for the general public and other stakeholders. This is accomplished in two ways: (1) full profiles of farmed species, where over 40 criteria and sub-criteria on ethology, but also ecology, physiology and general biology are reviewed in-depth according to the scientific literature. Based on these findings, a series of recommendations are proposed to address identified welfare issues in aquaculture; (2) short profiles of farmed species, where a sharp evaluation of 10 critical criteria is performed, covering ecological, behavioural and physiological traits transversal to all fish species: home and depth range, migration, reproduction, aggregation, aggression, habitat type, stress, malformations and slaughter. In addition, sustainable feeding and domestication are also addressed. These 10 criteria are answered using a welfare assessment protocol based on standardised risk analysis methods. For each species, this protocol provides (i) a comparison between the wild behaviour and the fish welfare state under conventional farming conditions, (ii) the overall welfare potential and (iii) the certainty of our findings. The sum of high scores of each species in these three measures throughout all 10 criteria results in the FishEthoScore, an index that summarises the general welfare state of the species. As occurs with all indexes, the FishEthoScore incurs the risk of oversimplifying a complex array of data concerning welfare. However, by asking the same questions to (ultimately) all farmed fish species, it offers a unique possibility not only for a comparative approach, but also for a global perspective on which species may be farmed most humanely. This is apparently the first effort to create such a welfare assessment scheme for aquaculture, and it may constitute a decisive step for a near-future welfare certification in the industry.

5. Fish domestication and the Five Freedoms

The concept of the Five Freedoms was coined in 1965 in the Brambell Report [112] concerning husbandry of livestock and revised by the Farm Welfare Council of the UK in 1979 into its present form [113]:

(I) Freedom from hunger and thirst—by ready access to fresh water and a diet to maintain full health and vigour;

(II) Freedom from discomfort—by providing an appropriate environment including shelter and a comfortable resting area;

(III) Freedom from pain, injury or disease—by prevention or rapid diagnosis and treatment;

(IV) Freedom to express normal behaviour—by providing sufficient space, proper facilities and company of the animal's own kind;

(V) Freedom from fear and distress—by ensuring conditions and treatment that avoid mental suffering.

This concept forms the basis of recommendations and legislations worldwide and, despite their age, are extensively employed for academic, educational and veterinary purposes with great practical utility [113]. Also, they paved the way so that animals could be considered by European law as sentient beings in the Lisbon Treaty of 2007 [114].

Research on animal sentience, situated between ethology and psychology, was initiated by ethologists such as Dawkins [115, 116] and cognitive psychologists such as Toates [117], around the early 1980's. These studies allowed a deeper understanding of animal minds and depicted how animals perceive the world and how environmental stimuli may affect their welfare level. Most importantly, they opened the door to the understanding of animal's subjective experiences. These experiences may be positive and negative, and the latter include suffering, which is fundamentally a wide range of unpleasant emotional (or emotion-like) states. Negative experiences occur when unpleasant subjective feelings are acute or continue for a long time when an animal is unable to carry out the actions that would normally reduce risks to life and reproduction in those circumstances [115, 118].

The five freedoms concept is nonetheless criticised. Some authors claim that this framework is overly guided by anthropocentric thinking about how animals ought to be handled, neglects the concept of allostasis (stability through change) and generally reflects a more ethical view than a science-based approach [119]. The concept may be misleading on, for example, stress and stress indicators such as cortisol [120], because it is well known that stress hormones are also involved in healthy adaptation [121–123]. Capacity to change, allostasis and biologically relevant challenges are crucial for good health and welfare, therefore stable conditions and homeostasis should not be considered optimal [119, 124]. As discussed in Section 3, phenotypes that are selected exclusively for production traits often show signs of structural and morphological imbalance. This occurs because symmorphosis (i.e. a match between structural design and functional demand) is disrupted [125]. In many farm animals (e.g. broiler chickens or heavily selected strains of trout), the structural design of internal organs does not match functional demand [48, 126]. This imbalance is responsible for many health problems in farm animals.

The Concept of Animal Welfare based on Allostasis aims to be an alternative to the Five Freedoms. It incorporates recent scientific developments in behavioural physiology and neurobiology and can be summarised as follows [119]:

- Stability through change (allostasis) and capacity to change are crucial to good health and good animal welfare. Health in this concept has the same meaning as defined in the World Health Organisation's (WHO) constitution as 'a state of complete physical, mental and social well-being and not merely the absence of disease or infirmity' [127].

- Good animal welfare is characterised by a broad predictive physiological and behavioural capacity to anticipate environmental challenges.
- Good animal welfare is guaranteed when the regulatory range of allostatic mechanisms matches the environmental demands.
- A low allostatic load (not very low or zero) is key for good health and good animal welfare.
- Symmorphosis should be respected.
- Behaviour and physiology should be interpreted in terms of animal perceptions and not exclusively in terms of human values.

To summarise, the Five Freedoms were primarily derived in relation to the welfare of farm animals, but, with the exception of the fifth freedom, would appear to consider that animals are passive within their environment [128]. Despite its undeniable role in the development of present (and future) welfare standards, this concept would benefit from an update in order to incorporate both ultimate (i.e. adaptive) and proximate (i.e. physiological) mechanisms. Integrating phylogeny and ontogeny in the design and analysis of husbandry practices would result in broader and overall better welfare schemes (**Figure 3**).

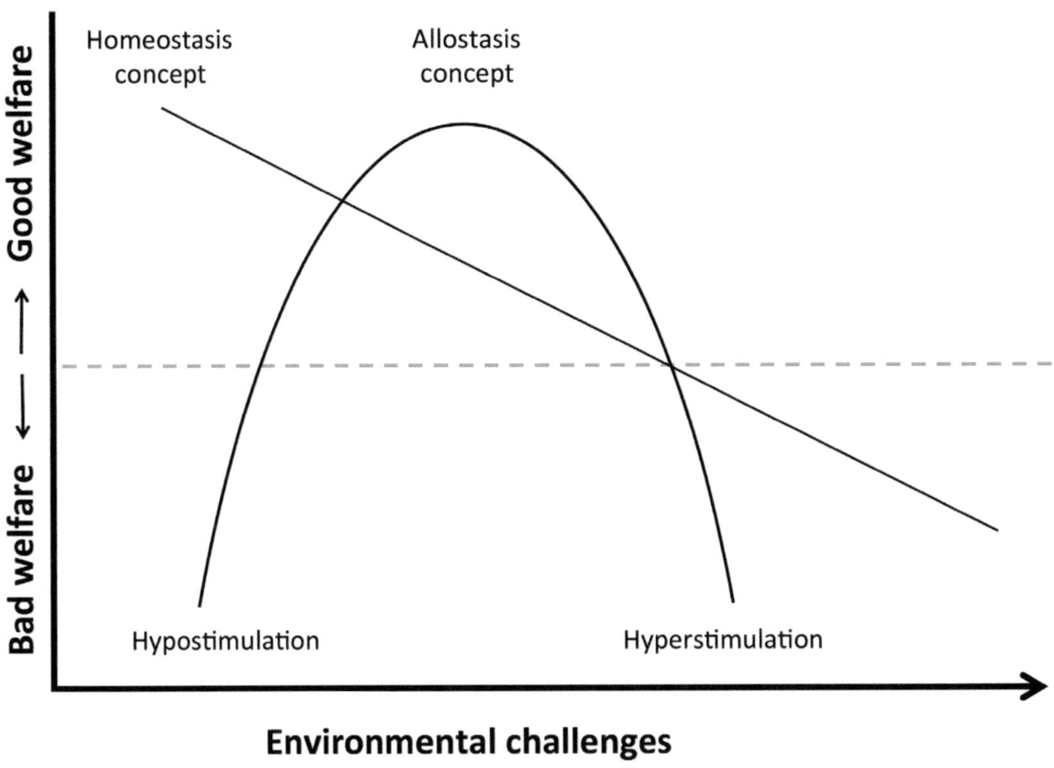

Figure 3. Animal welfare in relation to environmental challenges as shown by the out-dated concept based on homeostasis and the new concept based on allostasis. Adapted from [117].

6. Ethics vs. welfare in fish domestication

The ethical discussion on welfare of animals is controversial. This occurs because often the perspectives of scientists studying welfare science(s) and philosophers debating about ethics lie on very different standpoints. While science uses mostly operational and measurable concepts, such as the ones described throughout this chapter, ethics is focused on experiencing values and critically reflecting on them. Three main ethical theories are followed:

- The contract theory states that, despite the naturally selfish nature of man, there is a need to waiver rights in exchange for the benefits of a contract (implicit or explicit) [129];
- The utilitarianism theory states that the best action is the one that maximises utility, and therefore actions are to be evaluated majorly for their consequences [129];
- The deontologist theory states that the ethical value of an action is dependent on the motivation behind it and not on the consequences. It criticises the utilitarianism theory stating that the right of the individual can never be sacrificed for the common good [130].

The following allegory provides a good metaphor for the misunderstanding between philosophers and scientists:

'Two dog owners met one day to walk their dogs together. One owner had grown up in a small family that valued health, safety, and orderly, disciplined behaviour. The dog of this owner received regular veterinary care, two meals a day of low-fat dog food, and was walked on a leash. The other owner had grown up in a large community that valued conviviality, sharing of resources and close contact with the natural world. This dog (the owner's third - the first two had been killed by cars) had burrs in its coat, was fed generously but sporadically, and had never worn a collar in its life. Each owner, judging quality of life from very different viewpoints, felt sorry for the other's dog' [131].

The challenge lies in the different concepts, assumptions and vocabulary that scientists and philosophers use, which function as two distinct cultures with little mutual understanding or communication. Since the early days of the animal welfare debate, the two sides have struggled to communicate with each other, even though both were (and are) working with the common goal of understanding and improving an appropriate relationship between humans and other species of animals [132]. In fact, scientific research on animal welfare began because of ethical concerns over the quality of life of animals, and the public looks to animal welfare research for guidance regarding these concerns. The conception of animal welfare used by scientists must therefore relate to these ethical concerns in order to make sure that the orientation of the research and the interpretation of the findings are to address them successfully [131]. In order to bridge the gap and seek common ground between ethics and welfare science, it is important to recognise three classes of problems that may arise when the adaptations present in an animal do not fully correspond to the challenges posed by its current environment. These problems summarise the ethical concerns about the quality of life of animals [131]:

- If animals present adaptations that no longer serve a significant function in the new environment, then unpleasant subjective experiences may arise, yet these may not be accompanied by significant disruption of biological functioning. For example, species such as sea bream farmed in an open water sea cage may experience a strong, not full-filled need to seek shelter or forage on the sea bottom;

- If the environment poses challenges for which the animal has no corresponding adaptation, then functional problems may arise, even if not accompanied by significant effects on emotional-like states. Thus, a fish being fed with feed with incorrect lipid content will accumulate unhealthy body fat without appearing to notice or mind the problem;
- Where animals have adaptations corresponding to the kinds of environmental challenges they face, problems may still arise if the adaptations prove inadequate. For example, tilapia farmed in too cold water or trout farmed in too warm water will not be able to adequately regulate temperature, leading to functional failure as well as to a negative mental experience.

Animal welfare science has grown more compatible with the approaches used by some ethicists. Some scientists have recognised the interplay of normative and empirical elements in the assessment of animal welfare, and many are attempting to understand ethically relevant subjective experiences of animals. This convergence of the scientific and philosophical approaches may lead to a more integrated field of study and to a greater awareness that neither empirical information nor ethical reflection can, by themselves, answer questions about our proper relationship with animals of other species [132].

7. Conclusions

Considering that the domestication process in fishes is still in its early stages, determining whether and how this process affects welfare is not a straightforward task. Our understanding of fish biology is millennia behind that of terrestrial mammals, and the life-history of fish can be highly complex, with many species presenting stages that completely differ in every aspect from the final adult form. Furthermore, the sensory worlds of fish are very different from our own, and only recently have we begun to scratch the surface of the minds of fish, which hinders the establishment of empathy with our underwater relatives. To complicate things even more, fish farming is not focused on a few species, as in the case of land animals, but rather on hundreds of species that the industry invested in rearing for human consumption. Finally, the key concepts guiding welfare in farm animals are currently out-dated and seem to be insufficient to tackle a complex and diverse animal group such as fishes. The present review shows that domestication is not necessarily related to better welfare of fish especially because the traits the industry is selecting throughout the domestication process are generally focused on production (e.g. faster growth, larger mass), without taking into consideration pleiotropic or epistatic effects on other systems and on the organism. This knowledge gap should be bridged with research, either through species-specific approaches such as the COST action *Welfare of fish in European aquaculture* or broader frameworks such as FishEthoBase. Only by integrating the research and the industry, and by finding common

http://dx.doi.org/10.5772/intechopen.77251

grounds between ethics and science, can we expect to build species-appropriate and ethically justifiable systems in which to farm aquatic species.

Acknowledgements

This work was funded by Open Philanthropy Project (San Francisco, USA), Swiss Federal Food Safety and Veterinary Office (Bern, Switzerland), Stiftung Dreiklang (Basel, Switzerland), Haldimann-Stiftung (Aarau, Switzerland) and other private donations.

Author details

João L. Saraiva[1,2]*, Maria F. Castanheira[1], Pablo Arechavala-López[1,3], Jenny Volstorf[1] and Billo Heinzpeter Studer[1]

*Address all correspondence to: joao@fair-fish.net

1 Fair-Fish International, FishEthoBase Research Group, Winterthur, Switzerland

2 CCMAR - Center of Marine Sciences, Faro, Portugal

3 IMEDEA - Mediterranean Institute for Advanced Studies, Mallorca, Spain

References

[1] Price EO. Animal Domestication and Behavior. Wallingford, UK: Cabi; 2002

[2] Zeder MA, Hesse B. The initial domestication of goats (*Capra hircus*) in the Zagros Mountains 10,000 years ago. Science. 2000 Mar 24;**287**(5461):2254-2257

[3] Galal S. Biodiversity in goats. Small Ruminant Research. 2005 Oct 1;**60**(1):75-81

[4] Duarte CM, Marbá N, Holmer M. Rapid domestication of marine species. Science. 2007; **316**(5823):382

[5] Balon EK. About the oldest domesticates among fishes. Journal of Fish Biology. 2004 Dec 1;**65**:1-27

[6] El-Sayed A-FM. Tilapia Culture. Oxfordshire: CABI Publishing; 2006

[7] Teletchea F. Domestication of marine fish species: Update and perspectives. Journal of Marine Science and Engineering. 2015 Oct 15;**3**(4):1227-1243

[8] Teletchea F, Fontaine P. Levels of domestication in fish: Implications for the sustainable future of aquaculture. Fish and Fisheries. 2012;**15**(2):181-195

[9] Clutton-Brock J. The process of domestication. Mammal Review. 1992 Jun 1;**22**(2):79-85

[10] Lorenzen K, Beveridge MCM, Mangel M. Cultured fish: Integrative biology and management of domestication and interactions with wild fish. Biological Reviews. 2012 Aug 1;**87**(3):639-660

[11] Thorpe JE. Life history responses of fishes to culture. Journal of Fish Biology. 2004 Dec 1; **65**:263-285

[12] Arechavala-Lopez P, Fernandez-Jover D, Black KD, Ladoukakis E, Bayle-Sempere JT, Sanchez-Jerez P, et al. Differentiating the wild or farmed origin of Mediterranean fish: A review of tools for sea bream and sea bass. Reviews in Aquaculture. 2013;**5**:137-157

[13] Teletchea F. Domestication and genetics: What a comparison between land and aquatic species can bring? In Evolutionary Biology: Biodiversification from Genotype to Phenotype. Cham, Switzerland: Springer; 2015:389-401

[14] Gross MR. One species with two biologies: Atlantic salmon (*Salmo salar*) in the wild and in aquaculture. Canadian Journal of Fisheries and Aquatic Sciences. 1998 Jan 1;**55**(S1): 131-144

[15] Jensen P, Andersson L. Genomics meets ethology: A new route to understanding domestication, behavior, and sustainability in animal breeding. Ambio: A Journal of the Human Environment. 2005 Jun 1;**34**(4):320-324

[16] Martins CIM, Galhardo L, Noble C, Damsgård B, Spedicato MT, Zupa W, et al. Behavioural indicators of welfare in farmed fish. Fish Physiology and Biochemistry. 2012 Feb;**38**(1):17-41

[17] Duncan IJ. Science-based assessment of animal welfare: Farm animals. Revue scientifique et technique (International Office of Epizootics). 2005;**24**(2):483

[18] Duncan I, Dawkins M. The problem of assessing "well-being" and "suffering" in farm animals. In: Indicators Relevant to Farm Animal Welfare. Dordrecht, Netherlands: Springer; 1983. pp. 13-24

[19] Spruijt BM, van den Bos R, Pijlman FT. A concept of welfare based on reward evaluating mechanisms in the brain: Anticipatory behaviour as an indicator for the state of reward systems. Applied Animal Behaviour Science. 2001;**72**(2):145-171

[20] Broom DM. Indicators of poor welfare. The British Veterinary Journal. 1986 Nov 1;**142**(6): 524-526

[21] Broom DM. Animal welfare: Concepts and measurement. Journal of Animal Science. 1991;**69**(10):4167-4175

[22] Dawkins R. The Selfish Gene. Oxford, UK: Oxford University Press; 1976 497 p

[23] Krebs JR, Davies NB, Parr J. An Introduction to Behavioural Ecology. Oxford, UK: Blackwell Scientific Publications; 2012

[24] Wiens JJ. Polymorphism in systematics and comparative biology. Annual Review of Ecology and Systematics. 1999;**30**(1):327-362

[25] Kelly SA, Panhuis TM, Stoehr AM. Phenotypic plasticity: Molecular mechanisms and adaptive significance. Comprehensive Physiology. 2012;**2**:1417-1439. DOI: 10.1002/cphy.c110008

[26] Gjedrem T. Genetic improvement of cold-water fish species. Aquaculture Research. 2000 Jan;**31**(1, 1):25-33

[27] Duchev Z, Distl O, Groeneveld E. Early warning system for loss of diversity in European livestock breeds. Archiv Tierzucht. 2006;**49**(6):521

[28] Eknath A, Dey M, Rye M, Gjerde B, Abella TA, Sevilleja R, et al. Selective breeding of Nile tilapia for Asia. Vol. 27; 1998

[29] Haffray P, Tsigenopoulos C, Bonhomme F, Chatain B, Magoulas A, Rye M, et al. European sea bass-*Dicentrarchus labrax*. In: "Genetics of Domestication, Breeding and Enhancement of Performance of Fish and Shellfish", Viterbo, Italy, 12-17th June, 2006. 2007

[30] Janssen K, Chavanne H, Berentsen P, Komen H. Gilthead Seabream (*Sparus aurata*) – Current status of selective breeding in Europe. Abstract book of the International Symposium on Genetics in Aquaculture XII, Santiago de Compostela, Spain, June 21st–27th, 2015

[31] Harkness WJK, Dymond JR. The Lake Sturgeon: The History of its Fishery and Problems of Conservation. Fish & Wildlife Branch, Ontario Department of Lands and Forests, Toronto, Canada; 1961

[32] Hutchings JA. Adaptive phenotypic plasticity in brook trout, *Salvelinus fontinalis*, life histories. Écoscience. 1996 Jan;**3**(1, 1):25-32

[33] Johnston IA. Phenotypic plasticity of fish muscle to temperature change. In: Fish Ecophysiology. Dordrecht: Springer; 1993. pp. 322-340

[34] de MB, Mol J, Vigouroux R, Chaves P de T. Phenotypic plasticity in fish life-history traits in two neotropical reservoirs: Petit-Saut reservoir in French Guiana and Brokopondo reservoir in Suriname. Neotropical Ichthyology. 2009;**7**(4):683-692

[35] Saraiva J, Gonçalves D, Oliveira R. Ecological modulation of reproductive behaviour in the peacock blenny: A mini-review. Fish Physiology and Biochemistry. 2013;**39**(1):85-89

[36] Saraiva JL, Gonçalves DM, Oliveira RF. Environmental modulation of androgen levels and secondary sex characters in two populations of the peacock blenny *Salaria pavo*. Hormones and Behavior. 2010;**57**(2):192-197

[37] Saraiva JL, Gonçalves DM, Simões MG, Oliveira RF. Plasticity in reproductive behaviour in two populations of the peacock blenny. Behaviour. 2011;**148**(14):1457-1472

[38] Saraiva JL, Pignolo G, Gonçalves D, Oliveira RF. Interpopulational variation of the mating system in the peacock blenny *Salaria pavo*. Acta Ethologica. 2012;**15**(1):25-31

[39] Falconer DS. Early selection experiments. Annual Review of Genetics. 1992;**26**(1):1-16

[40] Hill WG. Caballero and A. Artificial selection experiments. Annual Review of Ecology and Systematics. 1992;**23**(1):287-310

[41] Rauw W, Kanis E, Noordhuizen-Stassen E, Grommers F. Undesirable side effects of selection for high production efficiency in farm animals: A review. Livestock Production Science. 1998;**56**(1):15-33

[42] Andersson L. Genetic dissection of phenotypic diversity in farm animals. Nature Reviews. Genetics. 2001 Feb 1;**2**:130

[43] Carlborg Ö, Kerje S, Schütz K, Jacobsson L, Jensen P, Andersson L. A global search reveals Epistatic interaction between QTL for early growth in the chicken. Genome Research. 2003 Mar 1;**13**(3):413-421

[44] Jensen P. Domestication, selection, behaviour and welfare of animals — Genetic mechanisms for rapid responses. Animal Welfare. 2010 May 1;**19**(2):7-9

[45] Fleming IA, Agustsson T, Finstad B, Johnsson JI, Björnsson BT. Effects of domestication on growth physiology and endocrinology of Atlantic salmon (*Salmo salar*). Canadian Journal of Fisheries and Aquatic Sciences. 2002 Aug 1;**59**(8):1323-1330

[46] Mason JW, Brynildson OM, Degurse PE. Comparative survival of wild and domestic strains of brook trout in streams. Transactions of the American Fisheries Society. 1967 Jul 1;**96**(3):313-319

[47] FishEthoBase Research team. FishEthoBase. Available from: http://fishethobase.net/en/ethology/

[48] Allen D, Rosenfeld J, Richards J. Physiological basis of metabolic trade-offs between growth and performance among different strains of rainbow trout. Canadian Journal of Fisheries and Aquatic Sciences. 2016 Apr 6;**73**(10):1493-1506

[49] Castanheira MF, Martins CI, Engrola S, Conceição LE. Daily oxygen consumption rhythms of Senegalese sole *Solea senegalensis* (Kaup, 1858) juveniles. Journal of Experimental Marine Biology and Ecology. 2011;**407**(1):1-5

[50] Benfey TJ. The physiology and behavior of triploid fishes. Reviews in Fisheries Science. 1999 Jan 1;**7**(1):39-67

[51] Scott MA, Dhillon RS, Schulte PM, Richards JG. Physiology and performance of wild and domestic strains of diploid and triploid rainbow trout (*Oncorhynchus mykiss*) in response to environmental challenges. Canadian Journal of Fisheries and Aquatic Sciences. 2014 Sep 18;**72**(1):125-134

[52] Cotter D, O'Donovan V, O'Maoiléidigh N, Rogan G, Roche N, Wilkins NP. An evaluation of the use of triploid Atlantic salmon (*Salmo salar* L.) in minimising the impact of escaped farmed salmon on wild populations. Aquaculture. 2000 Jun 1;**186**(1):61-75

[53] Koenig MK, Kozfkay JR, Meyer KA, Schill DJ. Performance of diploid and triploid rainbow trout stocked in Idaho Alpine Lakes. North American Journal of Fisheries Management. 2011 Mar 24;**31**(1):124-133

[54] Altimiras J, Axelsoon M, Claireaux G, Lefrançois C, Mercier C, Farrell AP. Cardiorespiratory status of triploid brown trout during swimming at two acclimation temperatures. Journal of Fish Biology. 2002 Jan;**60**(1, 1):102-116

[55] Hyndman CA, Kieffer JD, Benfey TJ. Physiology and survival of triploid brook trout following exhaustive exercise in warm water. Aquaculture. 2003 May 1;**221**(1):629-643

[56] Ojolick EJ, Cusack R, Benfey TJ, Kerr SR. Survival and growth of all-female diploid and triploid rainbow trout (*Oncorhynchus mykiss*) reared at chronic high temperature. Aquaculture. 1995 Apr 1;**131**(3):177-187

[57] Withler RE, Beacham TD, Solar II, Donaldson EM. Freshwater growth, smolting, and marine survival and growth of diploid and triploid coho salmon (*Oncorhynchus kisutch*). Aquaculture. 1995 Nov 1;**136**(1):91-107

[58] Woodward CC, Strange RJ. Physiological stress responses in wild and hatchery-reared rainbow trout. Transactions of the American Fisheries Society. 1987 Jul 1;**116**(4):574-579

[59] Wydoski RS, Wedemeyer GA, Nelson NC. Physiological response to hooking stress in hatchery and wild rainbow trout (*Salmo gairdneri*). Transactions of the American Fisheries Society. 1976 Sep 1;**105**(5):601-606

[60] Takahara T, Minamoto T, Doi H, Ito T, Kawabata Z. Differences between domesticated Eurasian and Japanese indigenous strains of the common carp (*Cyprinus carpio*) in cortisol release following acute stress. Ichthyological Research. 2014 Apr 1;**61**(2):165-168

[61] Barton BA. Stress in fishes: A diversity of responses with particular reference to changes in circulating corticosteroids. Integrative and Comparative Biology. 2002 Jul 1;**42**(3):517-525

[62] Lachance S, Magnan P. Performance of domestic, hybrid, and wild strains of brook trout, *Salvelinus fontinalis*, after stocking: The impact of intra- and interspecific competition. Canadian Journal of Fisheries and Aquatic Sciences. 1990 Dec 1;**47**(12):2278-2284

[63] Eknath AE, Tayamen MM, Palada-de Vera MS, Danting JC, Reyes RA, Dionisio EE, et al. Genetic improvement of farmed tilapias: The growth performance of eight strains of *Oreochromis niloticus* tested in different farm environments. Aquaculture. 1993 Apr 1; **111**(1):171-188

[64] Osure GO, Phelps RP. Evaluation of reproductive performance and early growth of four strains of Nile tilapia (*Oreochromis niloticus*, L) with different histories of domestication. Aquaculture. 2006 Mar 31;**253**(1):485-494

[65] Fraser TWK, Fjelldal PG, Hansen T, Mayer I. Welfare considerations of triploid fish. Reviews in Fisheries Science. 2012 Oct 1;**20**(4):192-211

[66] Price EO. Behavioral development in animals undergoing domestication. Applied Animal Behaviour Science. 1999 Dec 1;**65**(3):245-271

[67] Waples RS. Dispelling some myths about hatcheries. Fisheries. 1999 Feb 1;**24**(2):12-21

[68] Huntingford FA. Implications of domestication and rearing conditions for the behaviour of cultivated fishes. Journal of Fish Biology. 2004 Dec 1;**65**:122-142

[69] Huntingford F, Adams C. Behavioural syndromes in farmed fish: Implications for production and welfare. Behaviour. 2005 Sep 1;**142**(9–10):1207-1221

[70] Arechavala-Lopez P, Toledo-Guedes K, Izquierdo-Gomez D, Šegvić-Bubić T, Sanchez-Jerez P. Implications of sea bream and sea bass escapes for sustainable aquaculture management: A review of interactions, risks and consequences. Reviews in Fisheries Science & Aquaculture. 2018 Apr 3;**26**(2):214-234

[71] Kotrschal A, Rogell B, Bundsen A, Svensson B, Zajitschek S, Brännström I, et al. Artificial selection on relative brain size in the guppy reveals costs and benefits of evolving a larger brain. Current Biology. 2013;**23**(2):168-171

[72] Castanheira MF, Conceição Luís EC, Sandie M, Sonia R, Marie-Laure B, Børge D, et al. Coping styles in farmed fish: Consequences for aquaculture. Reviews in Aquaculture. 2017 Mar 27;**9**(1):23-41

[73] Gjedrem T, editor. Selection and Breeding Programs in Aquaculture. Dordrecht, Netherlands: Springer Netherlands; 2005

[74] Martins Catarina IM, Schrama Johan W, Verreth Johan AJ. Inherent variation in growth efficiency of African catfish *Clarias gariepinus* (Burchell, 1822) juveniles. Aquaculture Research. 2005 Jun 21;**36**(9):868-875

[75] Smiseth PT, Wright J, Kölliker M. Parent–offspring conflict and co-adaptation: Behavioural ecology meets quantitative genetics. Proceedings of the Royal Society of London - Series B: Biological Sciences. 2008 Aug 22;**275**(1645):1823-1830

[76] Dingemanse NJ, Kazem AJN, Réale D, Wright J. Behavioural reaction norms: Animal personality meets individual plasticity. Trends in Ecology & Evolution. 2010 Feb 1;**25**(2):81-89

[77] FAO. FAO Yearbook. Fishery and Aquaculture Statistics. 2015. Rome: FAO; 2017. p. 107

[78] Balcombe J. What a Fish Knows: The Inner Lives of our Underwater Cousins. Straus and Giroux: Farrar; 2016. p. 305

[79] Jordan DS. The history of ichthyology. Science. 1902;**16**(398):241-258

[80] Fay RR, Simmons AM. The sense of hearing in fishes and amphibians. In: Comparative Hearing: Fish and Amphibians. New York, NY: Springer; 1999. pp. 269-318

[81] Sneddon LU. Comparative physiology of nociception and pain. Physiology (Bethesda, Md.). 2018 Jan;**33**(1, 1):63-73

[82] Berthoz A. Neurobiology of 'Umwelt': How Living Beings Perceive the World. Springer Science & Business Media; 2008. p. 161

[83] Evans BIA. Fish's eye view of habitat change. In: The Senses of Fish. Dordrecht: Springer; 2004. pp. 1-30

[84] Northcutt RG, Wullimann MF. The visual system in teleost fishes: Morphological patterns and trends. In: Sensory Biology of Aquatic Animals [Internet]. New York, NY: Springer; 1988. pp. 515-552

[85] Hansen A, Reutter K. Chemosensory systems in fish: Structural, functional and ecological aspects. In: The Senses of Fish. Dordrecht: Springer; 2004. pp. 55-89

[86] Sorensen PW, Wisenden BD. Fish Pheromones and Related Cues. Ames, Iowa, USA: John Wiley & Sons; 2015. p. 305

[87] Keller-Costa T, Saraiva JL, Hubbard PC, Barata EN, Canário AVM. A multi-component pheromone in the urine of dominant male tilapia (*Oreochromis mossambicus*) reduces aggression in rivals. Journal of Chemical Ecology. 2016 Feb 1;**42**(2):173-182

[88] Saraiva JL, Keller-Costa T, Hubbard PC, Rato A, Canário AVM. Chemical diplomacy in male tilapia: Urinary signal increases sex hormone and decreases aggression. Scientific Reports. 2017 Aug 9;**7**(1):7636

[89] Valentinčič T. Taste and olfactory stimuli and behavior in fishes. In: The Senses of Fish. Dordrecht: Springer; 2004. pp. 90-108

[90] Ladich F. Sound production and acoustic communication. In: The Senses of Fish. Dordrecht: Springer; 2004. pp. 210-230

[91] Ashley PJ, Sneddon LU, McCrohan CR. Nociception in fish: Stimulus–response properties of receptors on the head of trout Oncorhynchus mykiss. Brain Research. 2007 Aug 29; **1166**(Supplement C):47-54

[92] Sneddon LU. Trigeminal somatosensory innervation of the head of a teleost fish with particular reference to nociception. Brain Research. 2003 May 16;**972**(1):44-52

[93] Baduy F, Soares J, Silva M, Canário AVM, Saraiva JL, Guerreiro PM. Critical thermal maximum and minimum in Australoheros facetus, a neo-tropical invader in the Iberian peninsula. FishMed. 2016;**12**:1-3

[94] Mogdans J, Kröther S, Engelmann J. Neurobiology of the fish lateral line: Adaptations for the detection of hydrodynamic stimuli in running water. In: The Senses of Fish. Dordrecht: Springer; 2004. pp. 265-287

[95] Janssen J. Lateral line sensory ecology. In: The Senses of Fish. Dordrecht: Springer; 2004. pp. 231-264

[96] Balcombe J. In praise of fishes: Précis of 'what a fish knows' (Balcombe 2016). Animal Sentience: An Interdisciplinary Journal on Animal Feeling. 2016 Jul 6;**1**(8):1

[97] Keller CH. Electroreception: Strategies for separation of signals from noise. In: The Senses of Fish. Dordrecht: Springer; 2004. pp. 330-361

[98] Scheffel A, Kramer B. Intra- and interspecific electrocommunication among sympatric mormyrids in the Upper Zambezi River. In: Ladich F, Collin SP, Moller P, Kapoor BG, editors. Communication in Fishes. Science Publishers, Enfield, NH, USA; 2006; pp. 733-751

[99] Tytell ED, Standen EM, Lauder GV. Escaping flatland: Three-dimensional kinematics and hydrodynamics of median fins in fishes. The Journal of Experimental Biology. 2008 Jan 15;**211**(2):187-195

[100] Power DM, Einarsdóttir IE, Pittman K, Sweeney GE, Hildahl J, Campinho MA, et al. The molecular and endocrine basis of flatfish metamorphosis. Reviews in Fisheries Science. 2008;**16**(sup1):95-111

[101] Pardo SA, Cooper AB, Dulvy NK. Avoiding fishy growth curves. Methods in Ecology and Evolution. 2013 Apr 1;**4**(4):353-360

[102] Randall DJ, Tsui TKN. Ammonia toxicity in fish. Marine Pollution Bulletin. 2002 Sep 1; **45**(1):17-23

[103] Ip YK, Chew SF. Ammonia production, excretion, toxicity, and Defense in fish: A review. Frontiers in Physiology. 2010 Oct 4;**1**:1-20

[104] Gross M. What fish genomes can tell us about life on land. Current Biology. 2013 May 20; **23**(10):R419-R421

[105] Vandepoele K, De Vos W, Taylor JS, Meyer A, Van de Peer Y. Major events in the genome evolution of vertebrates: Paranome age and size differ considerably between ray-finned fishes and land vertebrates. Proceedings of the National Academy of Sciences of the United States of America. 2004 Feb 10;**101**(6):1638-1643

[106] FAO. The State of World Fisheries and Aquaculture 2016. Contributing to food security and nutrition for all; 2016

[107] FAO. Domestic Animal Diversity Information System (DAD-IS). Available from: http://www.fao.org/dad-is/data/en/

[108] Ashley PJ. Fish welfare: Current issues in aquaculture. Applied Animal Behaviour Science. 2007 May 1;**104**(3):199-235

[109] Hastein T. Animal welfare issues relating to aquaculture. In: Proceedings of the Global Conference on Animal Welfare: An OIE Initiative; 2004. pp. 219-231

[110] Huntingford FA, Kadri S. Taking account of fish welfare: Lessons from aquaculture. Journal of Fish Biology. 2009 Dec 1;**75**(10):2862-2867

[111] COST AC 867. Welfare of Fish in European aquaculture; 2011

[112] Brambell FWR. Report of the Technical Committee to Enquire into the Welfare of Animals Kept under Intensive Livestock Husbandry Systems. London; 1965

[113] McCulloch SP. A critique of FAWC's five freedoms as a framework for the analysis of animal welfare. Journal of Agricultural and Environmental Ethics. 2013 Oct 1;**26**(5):959-975

[114] European Union. Treaty of Lisbon amending the treaty on European Union and the treaty establishing the European Community. Official Journal of the European Union. 2007

[115] Dawkins M. Animal Suffering: The Science of Animal Welfare. Dordrecht, Netherlands: Springer Netherlands; 1980

[116] Dawkins MS. Through our Eyes Only?: The Search for Animal Consciousness. Oxford, New York: Oxford University Press; 1998. p. 202

[117] Toates FM. Motivational Systems. CUP Archive; 1986. p. 204

[118] Carenzi C, Verga M. Animal welfare: Review of the scientific concept and definition. Italian Journal of Animal Science. 2009 Jan 1;**8**(Suppl1):21-30

[119] Korte SM, Olivier B, Koolhaas JM. A new animal welfare concept based on allostasis. Physiology & Behavior. 2007 Oct 22;**92**(3):422-428

[120] Webster AJ. What use is science to animal welfare? Die Naturwissenschaften. 1998 Jun;**85**(6):262-269

[121] McEwen BS. Protective and damaging effects of stress mediators. The New England Journal of Medicine. 1998 Jan 15;**338**(3):171-179

[122] Korte SM, Koolhaas JM, Wingfield JC, McEwen BS. The Darwinian concept of stress: Benefits of allostasis and costs of allostatic load and the trade-offs in health and disease. Neuroscience and Biobehavioral Reviews. 2005 Feb;**29**(1):3-38

[123] Koolhaas JM, Bartolomucci A, Buwalda B, de Boer SF, Flügge G, Korte SM, et al. Stress revisited: A critical evaluation of the stress concept. Neuroscience and Biobehavioral Reviews. 2011 Apr 1;**35**(5):1291-1301

[124] Schulkin J. Allostasis, Homeostasis, and the Costs of Physiological Adaptation. Cambridge, UK: Cambridge University Press; 2012. p. 386

[125] Weibel ER, Taylor CR, Hoppeler H. The concept of symmorphosis: A testable hypothesis of structure-function relationship. Proceedings of the National Academy of Sciences. 1991 Nov 15;**88**(22):10357-10361

[126] Korte SM, Sgoifo A, Ruesink W, Kwakernaak C, van Voorst S, Scheele CW, et al. High carbon dioxide tension (PCO2) and the incidence of cardiac arrhythmias in rapidly growing broiler chickens. The Veterinary Record. 1999 Jul 10;**145**(2):40-43

[127] World Health Organization. WHO | Constitution of WHO: Principles. WHO

[128] Ohl F, van der Staay FJ. Animal welfare: At the interface between science and society. Veterinary Journal. 2012 Apr;**192**(1, 1):13-19

[129] Burgess D. Utilitarianism, game theory and the social contract. Macalester Journal of Philosophy. 2005;**14**(1):7

[130] Alexander L, Moore M. Deontological ethics. In: Zalta EN, editor. The Stanford Encyclopedia of Philosophy. Metaphysics Research Lab, Stanford University, Stanford, USA; 2016

[131] Fraser D, Weary D, Pajor E, Milligan B. A scientific conception of animal welfare that reflects ethical concerns. Ethics: Animal ethics. 1997

[132] Fraser D. Animal ethics and animal welfare science: Bridging the two cultures. Applied Animal Behaviour Science. 1999 Dec 1;**65**(3):171-189

Chapter 7

Domestication of the Eurasian Perch (*Perca fluviatilis*)

Pascal Fontaine and Fabrice Teletchea

Additional information is available at the end of the chapter

http://dx.doi.org/10.5772/intechopen.85132

Abstract

The farming of percids (Eurasian perch *Perca fluviatilis*, pikeperch *Sander lucioperca*) has progressively become a diversification path of European inland aquaculture in the past 25 years. This required the domestication of wild or pseudowild (coming from polyculture ponds) populations. Considering the history of Eurasian perch, this domestication can be subdivided into four main successive parts: (1) a short initial prospective period (bibliographical analysis, market analysis, etc.), (2) a first experimental period to acquire basic data that notably resulted in the choice of the rearing system and commercial feeds, (3) a second experimental period allowing to get an in-depth knowledge on each of the main phase of the life cycle of this species (control of the life cycle in rearing conditions), and (4) a third experimental period, still ongoing, of optimization of rearing practices. This chapter allows understanding the domestication framework of this species and better understanding the role of different actors in the decision-making. In the future, the farming of this species is likely to rely on a larger diversity of rearing systems; a key issue is to study the interactions between species-rearing system. How different domestication trajectories or paths (intratrajectories variability) will affect global performances of Eurasian perch remains an open question.

Keywords: Eurasian perch, domestication, aquaculture, chronology, major steps, rearing system

1. Introduction

Fish farming is an animal production sector that followed, in the past years, various dynamic paths according to the region considered. For instance, between 1995 and 2015, this sector displayed a strong increase at global scale with a production rising from 14.9 to 51.3 Mt. (+242%), whereas only a slight increase was observed within the European Union countries: from 490,000 to 660,000 tons (+34%). At national level, fish production has decreased from 65,500 tons in 1995 to 44,500 tons in 2005 (−32%) in France, despite its expansion in other countries like Norway. This fact illustrates that the development of this sector depends strongly on territorial contexts. Despite projections indicating the strong increase of aquaculture at global scale up to 2050, much higher than any other animal production sectors, except for poultry production [1, 2], some territories are facing several obstacles. These obstacles include, among others, (i) competition with other economic sectors (fisheries, tourism, agriculture, production of potable water, etc.), for access to land and water resources, (ii) an economical context of free exchange that often results in strong competition with imported products coming from countries with much lower production costs, (iii) policies (environmental and social protection, food safety, etc.) most often perceived as very binding, and (iv) a degraded image of rearing systems and farming products for which sustainability is frequently questioned by societies in developed countries, particularly concerning quality of products, respect of animal welfare, and environmental impacts. All these issues could hamper the development of aquaculture in some developed countries, such as France. In this context, it is hard to conceive that fish farming could increase significantly in those regions. Nevertheless, these territories are heterogeneous and often display a strong historical, cultural (e.g., culinary traditions), and landscape (mountainous or coastal regions, ponds, wetlands, etc.) diversity that results in numerous microterritories with their specific consumption of fish or more exactly very typical products or dishes. This is particularly true for Europe and France. For instance, one might cite the consumption of smoked eel in the Netherlands [3], frying of cyprinids (roach, rudd) in the Valley of Moselle (Luxemburg), tench in the region of Extremadura in Spain, fried carps in the Sundgau in Alsace in France, meager in the southeast French Mediterranean Sea, or Eurasian perch in the countries around the Alps. These small markets rely on a close link between local populations, the history of the territory, and the presence of a specific landscape (e.g., country of ponds) or particular ecosystems (lakes) and the animal species inhabiting these regions. This tight link between consumers and species is obviously the case for the market of Eurasian perch in the Alps region, where consumers often require the presence of the fish skin to clearly observe the alternation of dark and light bands, typical of this species [4]. These are key advantages for this territory that could allow the development of a diversified and resilient aquaculture based on the diversification of the production and the domestication of new fish species, corresponding to a development model that we can call "mosaic aquaculture." This is in this global context associated with this vision that the domestication of Eurasian perch started in the early 1990s, 25 years ago. The understanding of the initial motivations and the process of domestication realized over this period require first considering the specificities of inland European aquaculture and associated territories.

In Europe (European Union), inland aquaculture only represents 25.3% of the total production [5]. Two main distinct economic sectors exist, salmoniculture (farming of salmonids, chiefly monoculture of rainbow trout (*Oncorhynchus mykiss*) in running waters and pond culture, corresponding to polyculture in ponds with the dominant species being common carp (*Cyprinus carpio*). Thus, logically, the two most consumed fish species in Europe are rainbow trout (second) and common carp (fifth), mainly in Central and Eastern Europe for the latter. The domestication of Eurasian perch started in France with the will to diversify inland aquaculture while respecting the other economic sectors already developed, particularly pond aquaculture. Interestingly, it is important to specify that in France, pond aquaculture is mainly for the

http://dx.doi.org/10.5772/intechopen.85132

restocking market in link with angling activities: fish are sold alive to river managers (associations of anglers) or private ponds. These markets are both more lucrative and less demanding in terms of personnel and investment. A very small percentage of this aquaculture production is destined to the markets for human consumption.

2. Why choosing Eurasian perch?

The initial choice of Eurasian perch resulted from several points that were taken into account locally, like at the Lorraine territory scale in France. First, at national level, there was at that time the mutual motivation by several stakeholders (producers, policymakers, and developing agencies) to promote and diversify freshwater aquaculture with different incentives, even though the human consumption market was targeted (**Table 1**). In Lorraine, this dynamism first resulted in one part in the structuring of the inter-profession with the establishment of the Inland Aquaculture Lorraine Sector (Filière Lorraine d'Aquaculture Continentale) in 1987 and on the other part the inception of a new specific university diploma in inland aquaculture, the "Ingénieur-Technologue" DI-T [6–8]. Besides, carnivorous fishes, such as Eurasian perch, pikeperch *Sander lucioperca*, or pike *Esox Lucius*, are and remain both the most appreciated species by anglers and consumers who know them, particularly in Western Europe (except salmonids). Third, a survey realized at the European scale revealed that in some territories (Eastern France, Switzerland, and Northern Italia), this species was widely consumed in various forms (whole fish, fillets, etc.) and at different sizes (**Table 2**) [9], and they exist a niche market relatively large such as in Switzerland where it was estimated at about 4000 tons of fillets per year with a supply essentially ensured by fisheries from large lakes in Central and Northern Europe and Russia [10–11]. Fourth, the production of Eurasian perch in polyculture ponds remains challenging to control, which is less the case for other carnivorous species. So much that in certain French regions (Centre), this species was considered as undesirable by fish farmers because of dwarfing problems often linked to the overabundance of young individuals [11].

Summing up, the domestication of Eurasian perch appeared as a good compromise for several reasons: (1) a diversification of aquaculture production targeting the human consumption market by valuing a native species known by consumers and benefiting from a good image

Species	Territories	Initial will	Current production in France
Black bass *Micropterus salmoides*	South-West	Angling, human consumption	Negligible
Siberian sturgeon *Acipenser baeri*	Aquitaine	To preserve another sturgeon species (*A. sturio*)	17 farms, third global producer of caviar
Eurasian perch *Perca fluviatilis*	Lorraine, Rhône-Alpes	Human consumption	100 tons, three perch farms
Wels *Silurus glanis*	Centre, Languedoc	Human consumption	Negligible

Table 1. Trials of diversification and domestication of new fish species in inland aquaculture in metropolitan France during the last decades of the twentieth century.

Countries	Production/exploited ecosystems	Valorization
Germany	Fisheries (large lakes, rivers, Baltic Sea)	Angling, exportation, weak human consumption
Austria	Fisheries (Constance Lake)	Exportation, human consumption
Belgium	Fisheries in rivers, polyculture in ponds	Angling
Bulgaria	Fisheries in rivers or in reservoirs	Angling, human consumption
Denmark	Fisheries in lakes or estuaries	Angling, exportation
Finland	Fisheries in Baltic Sea and inland waters	Angling, strong human consumption
France	Fisheries in lakes and rivers	Angling, strong human consumption (East)
Great Britain	Fisheries in lakes and rivers	Angling
Hungary	Fisheries in lakes and rivers	Angling
Ireland	Lough Neagh	Exportation
Luxemburg	Fisheries in rivers	Angling, weakly consumed
Norway	Fisheries in inland waters of East, South, and North-East	Angling, exportation, human consumption
Netherlands	Fisheries in IJsselmeer lakes and inland waters	Angling, weak human consumption
Baltic countries	Fisheries in lakes	Exportation
Poland	Fisheries in inland waters (Swinoujscie region)	Angling, exportation
Czech Republic and Slovakia	Fisheries in the Danube River and other rivers	Angling, human consumption
Romania	Fisheries in ponds, in the Danube River, Razelm Lake	Angling, human consumption
Serbia and Macedonia	Fisheries in the Danube River and lakes (Dojran Lake)	Human consumption
Sweden	Fisheries in the Baltic Sea	Angling, exportation, human consumption
Switzerland	Fisheries in lake	Angling, strong human consumption

Table 2. Interest for Eurasian perch according to European countries, survey realized in 1993 [9].

and an established market niche and (2) the development of a new activity that did not harm other traditional activities of the sector (no competition). Initially, this project of diversification aimed at developing a complementary activity for pond fish farmers. Besides, linking to the survey realized [9], a possible competition with capture fisheries coming from Eastern and Central Europe as well as Scandinavia was highlighted; yet, surveyed persons stated that the capture levels were highly variable from one year to another, product quality (filleting yield) also strongly varied (effect of reproductive cycle), and supply period of market was stopped during the spawning season in spring. Consequently, all these facts confirmed the possibilities to develop an aquaculture of Eurasian perch targeting a regular production of fresh fillets with a constant and high quality.

3. Acquiring knowledge on the biology of *P. fluviatilis* and *P. flavescens*

A the end of the 1980s and beginning of the 1990s, an in-depth analysis of the available literature on the biology of Eurasian perch and a North American close species, the yellow perch *P. flavescens*, was performed to better evaluate potentialities of this species. We first analyzed general articles as well as book chapters [12–22]. Then, we considered more specific studies focusing on the characteristics of populations inhabiting particular aquatic areas [13–27]. In the meantime, because some farming trials were already performed on yellow perch in the United States (large lake areas), a similar approach was realized aiming at establishing a synthesis of knowledge acquired on the zootechny of this sister species [28–38]. At this period, yellow perch was considered as the reference to promote the farming of Eurasian perch. This choice was reinforced by the fact that questioning about the rearing systems (ponds or recirculated systems) was similar. Based on these bibliographical analyses, preliminary thoughts resulted in the emergence of farming possibilities in Europe [39], and perciculture (i.e., farming of perch) was proposed as a possible way to diversity inland aquaculture in Europe [40].

3.1. Study of the life cycle of perch in natural conditions, first zootechnical trials, and choice of the rearing system

During the 1990s, researches were undertaken to first better know the life cycle of the species in local aquatic ecosystems, mainly in the Mirgenbach reservoir and Lindre ponds (Moselle, France), and second to determine the potential of this species at different stages (larval rearing, on-growing). The choice of the Mirgenbach was linked to the fact that this reservoir presents heated waters due to the nuclear power plant of Cattenom and could potentially present thermic conditions more favorable for the growth of perch, in the perspective of a future economic development. These field studies allowed describing the feeding regime, growth (relation size-weight), composition of the main tissues (muscles, gonads, liver, viscera), as well as the reproductive cycle [27, 41–44]. These data constituted the frame of reference and brought the basis for future experimentations, such as the control of the reproductive cycle. In parallel to these descriptive studies, first trials of acclimatization were realized using perch sampled at different development stages in natural conditions (e.g., egg ribbons mainly from the Leman Lake, INRA Thonon-les-Bains, Haute-Savoie, France), polyculture ponds (young perch of 4–20 g for Lorraine fish farm ponds), or rivers (eggs ribbons from Meuse). The acclimatization of young perch, either juveniles or sexually mature individuals, with diverse features from one year to another, was closely linked to the will to value stocks of fish often very abundant during fall and spring pond fisheries and displaying a low market value. Based on the works performed on the yellow perch [32, 34, 36], several weaning protocols were tested using feeds or diverse raw materials (beef liver, frozen plankton, dried or hydrated formulated feeds) [45]. Because of (i) very high mortality rate (40–60% in 2 months) linked to food refusal, development of pathologies caused by *Aeromonas hydrophila* and cannibalism, (ii) high variability of qualities of the different batches of fishes received (juveniles or mature fishes, sizes, more or less lean fish, etc.), and (iii) difficulty of weaning protocols, this way of developing perciculture was rapidly stopped. Nevertheless, it was maintained during few years to produce the biological material to realize growth trials and produce breeders [46]. This work allowed conducting a thinking on the choice of the rearing system, which was the most adapted to perciculture. If the production of juveniles could be realized in small ponds following extensive or semi-intensive methods [47], the on-growing phase was rapidly focused on rearing systems in controlled conditions, which allow higher production levels and a rationalization of rearing conditions to guarantee a reproducibility of performances and the development of the sector. Thus, on-growing trials were performed in floating cages (Lindre ponds, Lake of Féronval) and in recirculated aquaculture system (RAS) in Belgium and France. In this comparative approach of the possible potentialities by different rearing systems, it was demonstrated that similar specific growth rates were obtained in cages and RAS, but survival rates, feed conversion rates, and the homogeneity of individual weights were better in RAS [45, 48, 49]. It also appeared that perches farmed in cages had started a reproductive cycle: females and males captured in September (40–70 g) displayed gonadosomatic indexes of 2.4 and 7.1%, respectively, whereas they were constant and low in RAS (<0.5, sexual resting) [45, 48]. Yet, the development of gonads at such a low weight, lower than the market weight targeted (80–120 g), constituted a problem for maintaining optimal growth performances. These zootechnical trials also demonstrated that this species was very sensitive to pathogens, among which are parasites such as *Heteropolaria sp.*, a protozoaire [50, 51], or bacteria, such as *Aeromonas sobria* [52]. This sensitivity of this species led to the shutdown of the project of the enterprise Perlac SA located in the Lake Neuchâtel in Switzerland. The sensitivity of this species to external parasites, such as *Dactylogyrus* or *Costias*, was confirmed during the first rearing trials performed by the society Lucas Perches created in 2001 in France [53]. At this period, this society used the water from a small river "La petite seille" to decrease the water temperature coming from a geothermal forage used by the society. At last, a strong individual growth heterogeneity was observed during trials [50]. All these experiences realized in Belgium, France, and Switzerland resulted in the choice of RAS as the most adapted rearing system for the development of perciculture [54, 55]. This choice was confirmed by technical choices operated by the first perch farms, Percitech in Switzerland (society created in 1994) and Lucas Perches in France (created in 2002) (**Figure 1**). Since then, researches exclusively focus on this rearing system using diets for trout or sea bass mainly.

3.2. Control of the life cycle of Eurasian perch for the development of perciculture in RAS

Once the rearing system selected (intensive monoculture in RAS for the production of fillet for human consumption), diverse researches were performed in order to control the life cycle of the species in indoor conditions. They include the control of the reproductive cycle, the development of larval rearing protocols, the determination of nutritional needs, the optimization of growth performances, the control of quality of products, and first trials of genetic improvement. These researches were funded by both national (mainly in Belgium and France) and international, chiefly thanks to the European Union (FAIR-CT96-1572 1996-1998, FAIR-CT98-9241 1998-1999, Σ! 2321 ACRAPEP/ANVAR A0011134L 2001-2004, COOP-CT-2004-512629-PERCATECH 2004-2006) programs.

http://dx.doi.org/10.5772/intechopen.85132

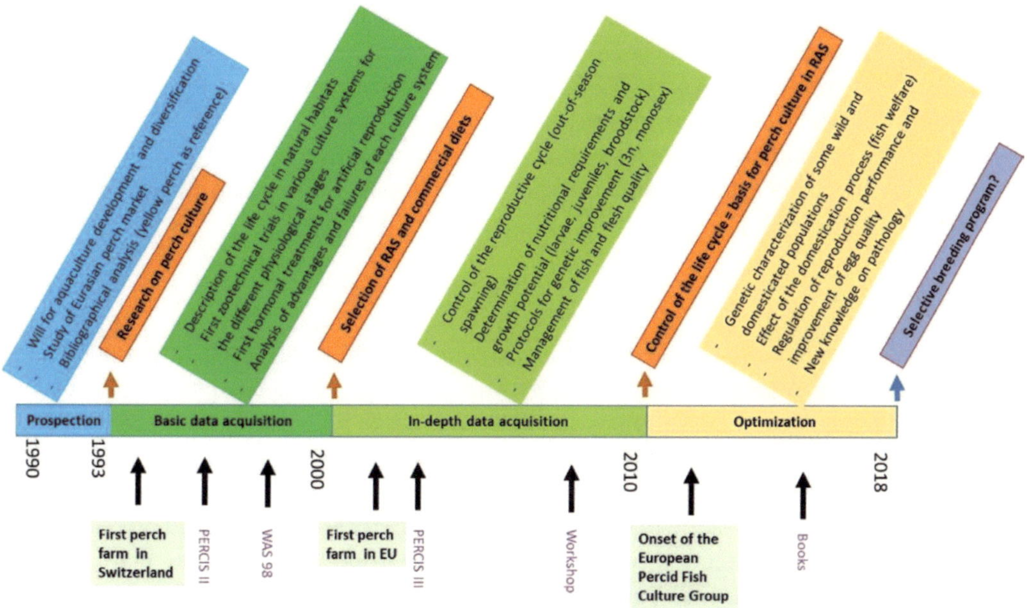

Figure 1. Timeline displaying the key phases of the domestication of Eurasian perch with from one hand the main knowledge acquired and the decisive decision taken (above the bar) and from the other hand the major events that occurred (below the bar) over the period 1990–2018.

3.3. Control of the reproduction

Even though the market for the perch fillet remains seasonal in the traditional consumption market (March–October), the development of an intensive monoculture in RAS required a complete control of the reproductive cycle in order to obtain out-of-season spawning and not only rely on the single annual reproduction occurring in spring [41–43, 56]. A first research axis focused on the environmental control of the reproductive cycle. A preliminary test demonstrated the possibility of controlling the reproductive cycle by manipulating both water temperature and duration of photoperiod [57]. Thereafter, these researches allowed disentangling the respective roles of water temperature variations and duration of photophase by distinguishing the different phases of a reproductive cycle: induction, vitellogenesis, and final steps of the cycle [58–66]. All these works allowed developing a reliable protocol for the induction of out-of-season spawning close to 100% [67]. This program is now routinely applied in farm conditions; it allowed the realization of 2–12 reproductive cycles per year with different batches of breeders managed in delayed conditions. If the temperature variations and the duration of photoperiod drive the timing of the successive steps of the reproductive cycle (determining factors), other factors can modulate the quality of reproductive performances observed. For example, the feeding strategy is very important, and thus the nutritional needs of breeders were specified [68, 69]. In fact, numerous rearing factors, including environmental, nutritional, and populational, can act on breeders and influence their reproductive performances; multifactorial approaches must be used to optimize rearing conditions and secure performances [70, 71].

Complementary to the control of reproductive cycle for the induction of out-of-season spawning, additional protocols based on hormonal injection were developed to synchronize spawning during the reproductive season [72–77]. They were based on previous works performed on the yellow perch [78–80]. The application of hormonal injections is now facilitated by the use of a classification method of oocyte stage maturation in preovulatory period [81]. At last, reliable protocols for collecting gametes (spermatozoa, oocytes) and artificial reproduction are also now available [82].

3.4. Larval rearing

Initially, trials of larval rearing were performed with spawning collected in various aquatic areas. Like for on-growing trials, several ways were initially prospected to promote the production of weaned juveniles: (1) an extensive production in small ponds with an *ex situ* weaning in tanks, (2) a semi-intensive production in mesocosms, and (3) an intensive production in RAS [83]. Even though few fish farms used the methods of mesocosms to produce the juveniles, particularly in Ireland, this is the intensive rearing in RAS that is mainly used nowadays. The first works aimed at optimizing the abiotic environment of farming (light intensity, duration of photophase, color of tank walls) and feeding protocols [84–87]. Initially, particular attention was paid to the use or not and the choice of live prey for larval rearing. The first protocols that have been developed used rotifers [88] or nauplii of *Artemia* spp. of various sizes [84, 87, 89–91]. The feeding transition (weaning = change from a feeding based on live prey to a commercial formulated diet) was soon questioned [92]. Very rapidly, major issues appeared: first, a high growth heterogeneity with a strong intra-cohort cannibalism rate [93–95] and second, the onset of developmental anomalies (malformations of skeleton and lordosis) with notably low inflation rate of the swim bladder [96–98].

The very strong impact of cannibalism within the first weeks of rearing was rapidly confirmed during the first commercial production [53]. Up to now, the strategy adopted by fish farmers to reduce cannibalism relies on frequent sorting (each week or 2 weeks) to maintain homogeneous batches during the nursery period and early weeks of on-growing. At that level, the results obtained by Mandiki et al. [99] suggested that they are natural populations less aggressive than others are, when they are placed in rearing conditions. Consequently it could be interesting to evaluate the intraspecific variability of wild populations (search for more docile populations). Concerning the problems of the inflation of the swim bladder and developmental anomalies often linked to the first point, they are mainly related to larval rearing conditions [100]. An improvement of rearing conditions associated with a high level of prophylaxis allowed increasing inflation rates and reduced malformation rates. In order to avoid the on-growing of individuals without swim bladder, protocols of sorting, based on practices realized in marine fish farming, were developed [101, 102]. Today, perch farms with well-conceived and seriously managed hatchery-nursery produce regular batches of 0.5 up to 1 million of weaned juveniles. However, developmental anomalies remain regularly observed in farms [103]. It is important to specify that the publication of a developmental table for the embryo-larvae corresponding to a normal development constitutes a major tool to identify the causes of common developmental anomalies) [104].

3.5. On-growing, nutritional needs

Once fry were available, trials on pre-on-growing and on-growing were realized in order to determine from one part the optimal conditions of growth and on the other part the potential of this species. It was first demonstrated that this species has a diurnal feeding activity [105]; the application of photoperiod with a long photophase stimulates growth and inhibits gonadal development [106]. First rearing trials had also demonstrated the gregarious behavior of this species (schooling behavior) and its ability to feed on pellets [46]. At this period, feeds for rainbow trout or sea bass were distributed to perch; feed conversion rates of 1.0–1.5 were registered according to the ration rate applied [49, 106–108]. High survival rates were also obtained (>80%).

Once these favorable prerequisites were established (gregarious behavior, sufficient survival, acceptability of artificial feeds, correct alimentary conversion rate, etc.), more dedicated researches were realized on the effects of both major abiotic and biotic factors on growth. Thus, it was demonstrated that the optimal temperature for growth was 22–24°C [107]. Thereafter, complementary works allowed specifying the effects of the rearing environment (tank wall color, light intensity, manipulations) on the ingested feed and growth [109–110]. The effects of rearing conditions on the physiological state of fish were also studied; perch appeared as very sensitive to both poor conditions and manipulations [111, 112]. At the feeding level, ration table for maintenance and optimal and maximal growth according to physiological stages were determined [107, 108, 113, 114]. Then, nutritional needs were progressively determined to promote the emergence of a feed for percids once the volume of production would be large enough. Thus, the nutritional requirements in proteins, lipids, and some additives, such as oxidative as ethoxyquin, were specified [89, 115–118]. These studies allowed defining that a feed for perch should contain 43–50% of proteins, 13–18% of lipids, and 10–15% of glucids [119].

3.6. Quality of products

The domestication of species for the human consumption market requires knowing and controlling the quality of products (whole fish, fillet). Thus, very early, once the first zootechnical trials were completed, the chemical composition of the tissue of perch, and notably muscle, was analyzed [41, 120]. One major goal was the production of constant quality fillet to consumers, similar to the wild fillet coming from the lake. Researches were started from one part to understand the natural variability of organoleptic properties of the perch fillet according to the origin of captures and, on the other part, to identify the determinants of this quality. Importantly, the quality of a product is a vague and complex notion that depends on nutritional, technological, sensorial, and sanitary features. Thus, features of perch coming from different regions (Geneva Lake, Rhine estuary) were compared among themselves and to perch obtained from RAS [121, 122]. It was found that first the quality of products was highly variable according to the natural environment studied and second that farming factors (feeds, rearing densities, etc.) strongly impacted the properties of farmed perch [123, 124]. In fact, the control of the quality of products (flesh or whole fish), over the course of domestication, is multifactorial [125, 126].

3.7. Manipulation of sex and ploidy: genetic management of domesticated populations

The Eurasian perch displays a sexual dimorphism of growth in favor of females [107, 108]; thus, the production of monosex female populations has rapidly appeared as a solution to reduce growth heterogeneity and increase growth performances. Hence, protocols (hormonal treatment with 17α-methyltestosterone) were developed for the production of homogametic males or neomales (XX) [127], with a sperm quality similar to heterogametic males [128]. Once produced and mature, those neomales were breeded with normal females (XX) allowing the production of 100% females, for which growth improvements were observed after 7 months of rearing in RAS at 23°C [129]. In a complementary study, trials of production of 100% female populations were also realized by gynogenesis using spermatozoa inactivated by UV radiation [130]. However, due to the low survival rates as well as insufficient growth performances, this method is rarely used [129].

A second path, triploidization, was also studied in order to produce sterile animals. This path also appeared as very important because Eurasian perch is a species that can start a reproductive cycle before reaching market size. It is possible to capture in natural habitats (ponds) sexually mature females and males as such low weights as 10–20 g, even lower for males. As for other species reared in fish farming (salmonids), protocols based on thermal or pressure chocks were also developed to produce triploid perch [131].

With the development of perch farms (7–8 farms localized in Germany, France, Ireland, and Switzerland) and the increase of production in RAS (estimated between 500 and 800 tons per year), first thinking on the necessity to develop selective breeding programs emerged, mainly to improve growth performances and decrease production costs. Yet, up to now, no true selective breeding programs exist, even though basic genetic knowledge was acquired to develop them. Studies have notably allowed to characterize the genetic variability of wild perch, very often used as founding populations of current farmed stocks [132–133] and stocks of domesticated breeders currently present in perch farms [134]. These studies have demonstrated that the available stocks of domesticated perch in farms were (i) sufficiently genetically variable to allow developing selective genetic programs (lack of consanguinity) and (ii) often genetically distant from the origin populations (Alpine lakes) presumably assumed by fish farmers.

4. Dissemination and knowledge transfer

The domestication of a species requires the onset of periods of exchanges between all stakeholders of the sector (**Figure 1**), notably to allow transfer of expertise and co-elaboration of projects based on the identification of priorities and major bottlenecks. Concerning Eurasian perch, very rapidly, the few research laboratories implied in this species cooperated and organized scientific seminars at different scales to allow sharing new knowledge. The meetings organized at the transatlantic level (Canada, USA, and Europe) aimed first at sharing works performed on Eurasian and yellow perch. Some of these events (Namur, 2008; Nancy, 2014) had for main objective exchanges between the socioeconomic stakeholders of the sector (fish

http://dx.doi.org/10.5772/intechopen.85132

farmers, designer of fish farms, traders in aquatic products, etc.). Progressively, knowledge was compiled in more and more comprehensive book [135, 136]. Obviously, this diffusion of knowledge and co-construction also occurred at local, regional, and national scales. In France, for instance, an informal group of exchanges, entitled "National group of pond carnivorous fish," often met in the beginning of the 1990s to discuss experience on various species (Wels, pikeperch, black-bass, and perch) that were the subject of diversification [137–140]. At the regional level, in Lorraine, the " Filière Lorraine d'Aquaculture Continentale (FLAC)" supports diverse zootechnical trials and, therefore, actively contributes to the emergence of perch farms on this territory. Later a similar initiative was taken in other regions from other countries, like in Ireland [141].

5. Conclusion

The domestication of Eurasian perch was initially based on local issues (niche market, development of activities and jobs in rural environments). This domestication occurred in a few main steps: (1) socioeconomic analysis of the market, (2) first zootechnical trials and choice of the major rearing system (RAS), and (3) acquisition of in-depth knowledge on the successive stages of the production cycle (control of the reproductive cycle and reproduction, control of the larval rearing, on-growing, and quality of products) (**Figure 1**). It is important to highlight that the first two steps strongly considered the knowledge previously acquired on a close species, the yellow perch. Today, the Eurasian perch is considered at the level 4 of domestication, which means that the entire life cycle is closed in captivity without any wild inputs but no selective breeding programs is applied [142].

Even though the first experimental trials were initiated at the beginning of the 1990s, the first perch farm (SARL Lucas Perches) created within the European Union was located in 2002 as a pilot enterprise. Importantly in Switzerland, a perch farm, Percitech, was created much earlier in 1994. About 20 years later, numerous projects were launched, some with very high expectations (e.g., FjordFresh Holding S/A in Estonia), in numerous European countries; 10 of these enterprises truly developed a commercial activity. Today, most perch farms pursue their activities; only few, mainly in Ireland (country where perch is not consumed), have stopped their activity. The investors that initially believed in this species were not issued from the aquaculture sector and discovered it. Sometimes, it corresponds to industrials that succeeded in other sectors and wants to diversify their activities. This initial distance from the aquaculture sector constitutes one of the reasons of the slow development of perciculture. Learning requires time. Without doubt, the domestication of Eurasian perch was and remained a particular human adventure, where the link between the species and humans is visible at different levels and various forms.

In terms of perspective, one can expect that this young sector will pursue its development first based on current farms, whose economic viability remains to be demonstrated and second in link with the emergence of new projects and expansion of the market toward new consumers. This new development could imply the production of both pikeperch and perch within the same farms. To support this development, it is imperative to reduce production costs, high in RAS, and secure current stocks. The decrease of cost production will require in priority the onset of selective breeding programs and genetic improvements, a standardization and rationalization of rearing protocols (e.g., percid feeds, ration tables, etc.) and a reduction of investment levels for the development of new perch farms. For some, the development of a monoculture of perch in ponds could be the solution because it will allow a strong decrease in production cost. On the security side, it is important to (i) better know pathologies associated with this species, notably virus, among which some might represent a major risk for percids [143] and (ii) specify the effects of the domestication process on rearing performances of this species. As any other domesticated species, biological responses and performances of perch are modified by the domesticating environment specific to the rearing system chosen and associated rearing practices. Thus, preliminary results indicated that reproductive performances [144, 145] and its sensitivity to stress and immune system [146–150] depend on farm conditions. In the future, the domesticating context (**Figure 2**) could strongly vary according to local environment, which could lead fish farmer to choose different rearing systems and to target different markets. Once this main choice realized (context and domesticating direction fixed), secondary choices will define the trajectory of domestication that will result in physiological stage capture in nature, of the dynamic of transition (progressive or sharp) and cultural practices used, practices that could evolve over time with different dynamics. This complexity is reinforced by the fact that a population engaged in a specific domesticating context could change into another context because of a modification in the project, as was the case for perch reared in polyculture ponds, then weaned, and grown in cages or RAS (**Figure 2**). This diversity of directions and domesticating trajectories should lead to different evolutions (behavior, stress physiology, reproduction, etc.) variable from a context to another. These evolutions could even lead fish farmers to reconsider the initial choice of founding populations, given the enormous genetic diversity available in wild populations [4].

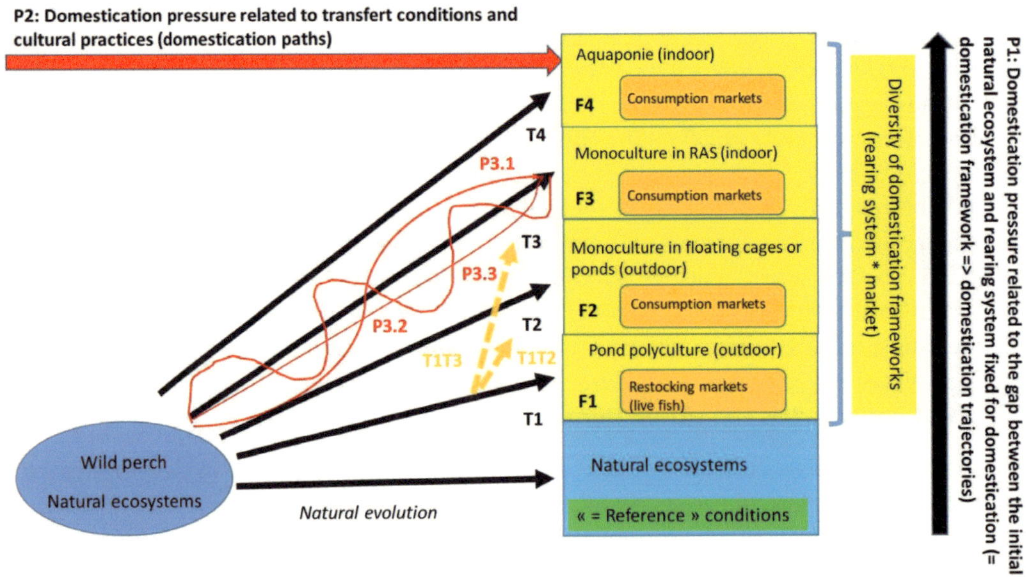

Figure 2. Diagram explaining the different domestication frameworks and pressures encountered by Eurasian perch during current farming trials (F: Framework, T: Trajectory, P: Path).

Author details

Pascal Fontaine* and Fabrice Teletchea

*Address all correspondence to: p.fontaine@univ-lorraine.fr

University of Lorraine, Nancy, France

References

[1] MacLeod M, Gerber P, Mottet A, Tempio G, Falcucci A, Opio C, et al. Greenhouse gas emissions from pig and chicken supply chains—A global life cycle assessment. In: Food and Agriculture Organization of the United Nations. Rome: FAO; 2013. p. 196

[2] FAO. La situation mondiale des pêches et de l'aquaculture 2018. Atteindre les objectifs de développement durable. Rome. 2018. p. 254

[3] FranceAgriMer. Etude d'initiatives françaises potentielles dans la filière européenne de l'anguille. VIA AQUA, Les études de FranceAgriMer. 2014. p. 15

[4] Stepien CA, Taxonomy HAE. Chap. 1. Distribution, and evolution of the percidae. In: Kestemont P, Dabrowski K, Summerfelt RC, editors. Biology and Culture of Percid Fishes—Principles and Practices. Dordrecht, The Netherlands: Springer; 2015. pp. 3-60

[5] Hough C editor. European Aquaculture Technology and Innovation Platform. A Review of the EATiP Strategic Research and Innovation Agenda. 2017. p. 29

[6] Condé B, Terver D. Filière Lorraine d'Aquaculture Continentale—Compte-rendu des journées du 5 juin et 16 septembre 1987. Supplément à la Revue française d'Aquariologie. 1998;**1**:39

[7] Fontaine P. Une région dans la cour des grands de l'aquaculture continentale. A l'est, du nouveau. Aqua Revue. 1993;**49**:25-28

[8] Terver D, Condé B, Fontaine P, Georges A. Aquariologie et aquaculture continentale à l'Aquarium tropical de Nancy. Objectifs culturels, pédagogiques et scientifiques: Bilan et perspectives. 2ème partie: Aquaculture. Supplément à la Revue française d'Aquariologie. 1993;**1**:8

[9] Tamazouzt L, Dubois JP, Fontaine P. Productions et marchés de la perche commune (*Perca fluviatilis*) en Europe. La Pisciculture Française. 1993;**114**:4-8

[10] Donzé F, Gradoux F. Frais du lac? L'illustré. 1998;**32**:24-27

[11] Fontaine P. Avancée récente des connaissances en perciculture. Aquafilia. 2004;**1**:5-8

[12] Turner CL. The seasonal cycle in the spermary of the perch. Journal of Morphology. 1919;**32**:681-711

[13] Le Cren ED. The length-weight relationship and seasonal cycle in gonad weight and condition in the perch (*Perca fluvitilis*). Journal of Animal Ecology. 1951;**20**:201-219

[14] Le Cren ED. Observations on the growth of perch (*Perca fluviatilis* L.) over twenty-two years with special reference to the effects of temperature and change in population density. Journal of Animal Ecology. 1958;**27**:287-233

[15] Mansuetti AJ. Early development of yellow perch, *Perca flavescens*. Chesapeake Science. 1964;**5**:46-66

[16] Jezierska B. The effect of various type of food on the growth and chemical composition of the body of perch (*Perca fluviatilis*) in laboratory conditions. Polish Archives Hydrobiology. 1974;**21**:467-479

[17] Collette BB, Ali MA, Hokanson KEF, Nagiec SA, Smirnov SA, Thorpe JE, et al. Biology of the percids. Journal of the Fisheries Research Board of Canada. 1977;**34**:1890-1899

[18] Thorpe JE. Synopsis of biological data on the perch *Perca fluviatilis* L. 1758 and *Perca flavescens* Mitchell 1814. FAO Fisheries Synopsis. 1977;**113**:139

[19] Hoestalndt M. Recherches biologiques sur la perche, *Perca fluviatilis* L., en France. Rapport Conseil Supérieur de la Pêche. 1979. p. 105

[20] Zelenkov VM. Early gametogenesis and sex differentiation in the perch, *Perca fluviatilis*. Journal of Ichthyology. 1981;**12**:124-130

[21] Craig JF. The Biology of Perch and Related Fish. Portland, Oregon: Timber Press; 1987. 333 p

[22] consumption KPF. Growth and recruitment in perch (*Perca fluviatilis* L.). [PhD thesis]. Sweden: University of Uppsala; 1987. p. 130

[23] Malservisi A, Magnin E. Changements cycliques annuels se produisant dans les ovaires de *Perca fluviatilis flavescens* (Mitchill) de la région de Montréal. Naturaliste Canadien. 1968;**95**:929-945

[24] Makarova NP. Seasonal changes in some of the physiological characteristics of the perch (*Perca fluviatilis*) of Ivan'kova reservoir. Journal of Ichthyology. 1973;**13**:742-752

[25] Craig JF. The body composition and of adult perch, *Perca fluviatilis* in Windermere. Journal of Animal Ecology. 1977;**46**:617-632

[26] Tanasichuk RW, Mackay WC. Quantitative and qualitative characteristics of somatic and gonadal growth of yellow perch *Perca flavescens*, from lac Ste Anne, Alberta. Canadian Journal of Aquatic Sciences. 1989;**46**:889-894

[27] Flesch A. Biologie de la perche (*Perca fluviatilis*) dans le réservoir du Mirgenbach (Cattenom, Moselle) [PhD thesis]. Université de Metz, France. 1994. p. 241

[28] Hale JG, Carlson AR. Culture of yellow perch in the laboratory. Progressive Fish Culturist. 1972;**34**:195-198

[29] Huh HT. Bioenergetics of food conversion and growth of yellow perch (*Perca flavescens*) and walleye (*Stizostedion vitreum vitreum*) using formulated diets [PhD thesis]. University of Wisconsin, USA. 1975. p. 196

[30] Hokanson KEF. Optimum culture requirements of the early life phases of the yellow perch. In: Soderberg RW editor. Perch Fingerling Production for Aquaculture. Wisconsin, USA: University of Wisconsin; 1978. pp. 24-40. Sea Grant Advisory Report #421 (p. 77)

[31] Kayes TB. Reproductive biology and artificial propagation methods for adult perch. In: Soderberg RW, editor. Perch Fingerling Production for Aquaculture. University of Wisconsin; 1978. pp. 6-23. Sea Grant Advisory Report # 421; p. 77

[32] Schott EF, Kayes TB, Calbert HE. Comparative growth of male versus female yellow perch fingerlings under controlled environmental conditions. American Fisheries Society Special Publications. 1978;**11**:181-186

[33] West G, Leonard J. Culture of yellow perch with emphasis on the development of eggs and fry. American Fisheries Society Special Publications. 1978;**11**:172-176

[34] Best CD. Initiation of artificial feeding and the control of sex differentiation in yellow perch, *Perca flavescens* [PhD thesis]. Madison, Wisconsin, USA: University of Wisconsin; 1981. p. 121

[35] Manci WE, Malison JA, Kayes TB, Kuszynski TE. Harvesting photopositive juvenile fish from a pond sing a lift net and light. Aquaculture. 1983;**34**:157-164

[36] Heidinger RC, Kayes TB. Yellow perch. In: Stickney RR, editor. Culture of Nonsalmonid Freshwater Species. Boca Raton, FL: CRC Press; 1986. pp. 103-113

[37] Hinshaw M. Factors affecting survival and growth of larval and early juvenile perch (*Perca flavescens*, Mitchill) [PhD thesis]. University of North Carolina State, USA. 1986. 80 p

[38] Malison JA, Held JA. Effects of fish size at harvest, initial stocking density and lighting conditions on the habituation of pond reared yellow perch (*Perca flavescens*) to intensive culture condition. Aquaculture. 1992;**104**:67-78

[39] Goubier V, Marchandise B. Etude de la faisabilité d'une pisciculture de perche. Rapp. Inst. Rég. Recher. Aqua. 1990. p. 70

[40] Fontaine P, Tamazouzt L, Terver D, Georges A. Actual state of production of perch: Problems and prospects. I—Mass rearing potentialities of the common perch under controlled conditions. In: Kestemont P, Billard R, editors. Aquaculture of Freshwater Species (Except Salmonid). Oostende, Belgium: European Aquaculture Society. 1993;**20**:46-48

[41] Sulistyo I. Contribution à l'étude et à la maîtrise du cycle de reproduction de la perche eurasienne *Perca fluviatilis* L [PhD thesis]. University Henri Poincaré, Nancy, France. 1998. p. 144

[42] Sulistyo I, Rinchard J, Fontaine P, Gardeur J-N, Capdeville B, Kestemont P. Reproductive cycle and plasma levels of sex steroids in female European perch *Perca fluviatilis*. Aquatic Living Resources. 1998;**11**:101-110

[43] Sulistyo I, Fontaine P, Rinchard J, Gardeur J-N, Migaud H, Capdeville B, et al. Reproductive cycle and plasma levels of sex steroids in male European perch *Perca fluviatilis*. Aquatic Living Resources. 1998;**13**:80-106

[44] Migaud H, Mandiki R, Gardeur J-N, Fostier A, Kestemont P, Fontaine P. Synthesis of sex steroids in final oocyte maturation and induced ovulation in female European perch, *Perca fluviatilis*. Aquatic Living Resources. 2003;**16**:380-388

[45] Tamazouzt L. L'alimentation artificielle de la perche *Perca fluviatilis* en milieux confinés (eau recyclée, cage flottante): Incidence sur la survie, la croissance et la composition corporelle. Thèse Université Henri Poincaré, Nancy, France. 1995. p. 126

[46] Fontaine P, Vlavonou R, Tamazouzt L, Terver D, Masson G. Essai d'élevage de perches sevrées en eau recyclée: Résultats préliminaires. Cahiers d'Ethologie Appliquée. 1994;**13**: 411-420

[47] Mélard C, Philippart JC. Essai d'élevage semi-intensif en bassins d'alevinage de perche fluviatile (*Perca fluviatilis* L.) obtenus par reproduction artificielle. Cahiers d'Ethologie Appliquée. 1984;**4**:59-66

[48] Fontaine P, Tamazouzt L, Capdeville B. Growth of the Eurasian perch *Perca fluviatilis* L. reared in floating cages and in water recirculated system: First results. Journal of Applied Ichthyology. 1996;**12**:181-184

[49] Tamazouzt L, Dubois JP, Fontaine P, Capdeville B, Terver D. Zootechnical performances and body composition of *Perca fluviatilis* fed pelleted diet in a floating cage: Effect of daily ration. Annales Zoologici Fennici. 1996;**33**:635-641

[50] Mélard C, Baras E, Kestemont P. Preliminary results of Eurasian perch (*Perca fluviatilis*) intensive rearing trials: effect of temperature and size grading on growth. Bulletin Français de la Pêche et de la Pisciculture. 1995;**336**:19-27

[51] Grignard JC, Mélard C, Baras E, Poirier A, Philippart JC, Bussers JC. Occurrence and impact of *Heteropolaria* sp. (Protozoa, Ciliophora) on intensively cultured perch (*Perca fluviatilis*). Annales Zoologici Fennici. 1996;**33**:653-657

[52] Chappuis V. Un élevage voué à l'échec? Terres & Nature. 2008;**5**:5

[53] Fontaine P, Ritter B. Premiers résultats en perciculture hors-sol chez Lucas Perches. Aquafilia. 2004;**2**:10-14

[54] Mélard C, Kestemont P. Diversification de la pisciculture en Région Wallonne: mise au point de l'élevage intensif de cyprinidés et percidés. Rapport de recherches à la Région Wallonne. Namur: Université de Liège, Facultés Universitaires Notre-Dame de la Paix; 1995. p. 58

[55] Fontaine P, Sulistyo I, Capdeville B, Kestemont P. Avancées récentes concernant la biologie et le contrôle de la reproduction de la perche eurasienne *Perca fluviatilis*. La Piscicuture Française. 1998;**133**:27-33

[56] Teletchea F, Fostier A, Kamler E, Gardeur J-N, Le Bail P-Y, Jalabert B, et al. Comparative analysis of reproductive traits in 65 freshwater fish species: Application to the domestication of new fish species. Reviews in Fish Biology and Fisheries. 2009;**19**:403-430

[57] Tamazouzt L, Fontaine P, Terver D. Décalage de la période de reproduction de la perche (*Perca fluviatilis*) en eau recyclée. Ichtyophysiologica Acta. 1994;**7**:29-40

[58] Migaud H, Fontaine P, Sulistyo I, Kestemont P, Gardeur J-N. Induction of out-of-season spawning in Eurasian perch *Perca fluviatilis*: Effects of cooling and chilling periods on female gametogenesis and spawning. Aquaculture. 2002;**205**:253-267

[59] Migaud H, Gardeur J-N, Kestemont P, Fontaine P. Off-season spawning of Eurasian perch *Perca fluviatilis*. Aquaculture International. 2004;**12**:87-102

[60] Migaud H, Fontaine P, Kestemont P, Wang N, Brun-Bellut J. Influence of photoperiod on the onset of gonadogenesis in Eurasian perch *Perca fluviatilis*. Aquaculture. 2004;**241**: 561-574

[61] Migaud H, Wang N, Gardeur J-N, Fontaine P. Influence of photoperiod on reproductive performances in Eurasian perch *Perca fluviatilis*. Aquaculture. 2006;**252**:385-393

[62] Fontaine P, Pereira C, Wang N, Marie M. Influence of pre-inductive photoperiod variations on Eurasian perch *Perca fluviatilis* broodstock response to an inductive photothermal program. Aquaculture. 2006;**255**:410-416

[63] Wang N, Gardeur J-N, Henrotte E, Marie M, Kestemont P, Fontaine P. Determinism of the induction of the reproductive cycle in female Eurasian perch, *Perca fluviatilis*: Effects of environmental cues and modulating factors. Aquaculture. 2006;**261**:706-714

[64] Wang N, Teletchea F, Kestemont P, Milla S, Fontaine P. Photothermal control of the reproductive cycle in temperate fishes. Reviews in Aquaculture. 2010;**2**:209-222

[65] Abdulfatah A, Fontaine P, Kestemont P, Gardeur J-N, Marie M. Effects of photothermal kinetic and amplitude of photoperiod decrease on the induction of the reproduction cycle in female Eurasian perch *Perca fluviatilis*. Aquaculture. 2011;**322-323**:169-176

[66] Abdulfatah A, Fontaine P, Kestemont P, Milla S, Marie M. Effects of the thermal threshold and the timing of temperature reduction on the initiation of reproduction cycle in female of Eurasian perch *Perca fluviatilis*. Aquaculture. 2013;**376-379**:90-96

[67] Fontaine P, Wang N, Hermelink B. Chapter 3. Broodstock management and control of the reproductive cycle. In: Kestemont P, Dabrowski K, Summerfelt RC, editors. Biology and Culture of Percid Fishes: Principles and Practices. Dordrecht, The Netherlands: Springer; 2015. pp. 103-122

[68] Henrotte E, Mandiki R, Prudencio A, Vandecan M, Mélard C, Kestemont P. Egg and larval quality, and egg fatty acid composition of Eurasian perch breeders (*Perca fluviatilis*) fed different dietary DHA/EPA/AA ratios. Aquaculture Research. 2010;**41**:51-63

[69] Henrotte E, Kaspar V, Rodina M, Psenicka M, Linhart O, Kestemont P. Dietary n-3/n-6 ratio affects the biochemical composition of Eurasian perch (*Perca fluviatilis*) semen but not indicators of sperm quality. Aquaculture Research. 2010;**41**:31-38

[70] Wang N. Déterminisme multifactoriel de la qualité du cycle de reproduction chez la perche commune *Perca fluviatilis*. Approche multifactorielle. Thèse de l'Université Henri Poincaré. Nancy. 2006;**1**:140 p

[71] Castets M-D. Fonction de reproduction et régulation de la qualité chez la perche commune, *Perca fluviatilis* [thèse de l'I.N.P.L]. Nancy. 2011. p. 301

[72] Kucharczyk D, Kujawa R, Mamcarz A, Skrzypczak A, Wyszomirska E. Induced spawning in perch, *Perca fluviatilis* L. using carp pituitary extract and HCG. Aquaculture Research. 1996;**27**:847-852

[73] Kucharczyk D, Kujawa R, Mamcarz A, Skrzypczak A, Wyszomirska E. Induced spawning in perch, *Perca fluviatilis* L. using FSH + LH with pimozide or metoclopramide. Aquaculture Research. 1998;**29**:131-136

[74] Kucharczyk D, Szczerbowski A, Luczynski MJ, Kujawa R, Mamcarz A, Wyszomirska E, et al. Artificial spawning of Eurasian perch, *Perca fluviatilis* L. using ovopel. Archives of Polish Fisheries. 2001;**9**:39-49

[75] Kouril J, Linhart O, Relot P. Induced spawning of perch by means of a GnRH analogue. Aquaculture International. 1997;**5**:375-377

[76] Rónyai A, Lengyel SA. Effects of hormonal treatments on induced tank spawning of Eurasian perch (*Perca fluviatilis* L.). Aquaculture International. 2010;**41**:345-347

[77] Targońska K, Szczerbowski A, Zarski D, Luczynski MJ, Szkudlarek M, Gomulka P, et al. Comparison of different spawning agents in artificial out-of-season spawning of Eurasian perch, *Perca fluviatilis* L. Aquaculture Research. 2014;**45**:765-767

[78] Goetz FW, Theofan G. In vitro stimulation of germinal vesicle breakdown and ovulation of yellow perch (*Perca flavescens*) oocytes. Effects of 17a-hydroxy-20ß-dihyroxyprogesterone and prostaglandins. General and Comparative Endocrinology. 1979;**37**:273-285

[79] Goetz FW, Duman P, Berndtson A, Janowsky EG. The role of prostaglandins in the control of ovulation in yellow perch, *Perca flavescens*. Fish Physiology and Biochemistry. 1989;**7**:163-168

[80] Dabrowski K, Ciereszko A, Ramseyer L, Culver D, Kestemont P. Effects of hormonal treatment on induced spermiation and ovulation in the yellow perch (*Perca flavescens*). Aquaculture. 1994;**120**:171-180

[81] Żarski D, Bokor Z, Kotrik L, Urbányi B, Horváth A, Targońska K, et al. A new classification of a preovulatory oocyte maturation stage suitable for the synchronization of ovulation in controlled reproduction of Eurasian perch, *Perca fuviatilis* L. Reproductive Biology. 2011;(3):194-209

[82] Żarski D, Horváth A, Kotrik L, Targońska K, Palińska K, Krejszeff S, et al. Effect of different activating solutions on the fertilization ability of Eurasian perch, *Perca fuviatilis* L., eggs. Journal of Applied Ichthyology. 2012;**28**:967-972

[83] Kestemont P, Mélard C, Held JA, Dabrowski K. Chapter 9. Culture methods of Eurasian perch and yellow perch early life stages. In: Kestemont P, Dabrowski K, Summerfelt RC, editors. Biology and Culture of Percid Fishes: Principles and Practices. Dordrecht, The Netherlands: Springer; 2015. pp. 265-293

[84] Fiogbé ED. Contribution à l'étude des besoins nutritionnels chez les larves et juvéniles de la perche fluviatile (*Perca fluviatilis* L.) [thèse de doctorat]. Facultés Universitaires Notre-Dame de la Paix, Namur, Belgique. 1996. p. 334

[85] Tamazouzt L, Leray C, Escaffre AM, Terver D. Effects of particle size on *Perca fluviatilis* larval growth. Aquatic Sciences. 1998;**60**:89-98

[86] Tamazouzt L, Chatain B, Fontaine P. Tank wall colour and light level affect growth and survival of perch larvae (*Perca fluviatilis* L.). Aquaculture. 2000;**182**:85-90

[87] Jourdan S. Influence de facteurs abiotiques, la photopériode et l'intensité lumineuse, sur la survie et la croissance de larves, post-larves et juvéniles de perche *Perca fluviatilis* L [thèse de l'I. N.P.L.]. Nancy. 1999. 147 p

[88] Awaiss A, Kestemont P, Micha JC. Nutritional suitability of the rotifer *Brachionus calyciflorus* Pallas for rearing freshwater fish larvae. Journal of Applied Ichthyology. 1992;**8**:263-270

[89] Kestemont P, Mélard C, Fiogbé ED, Vlavonou R, Masson G. Nutritional and animal husbandry aspects of rearing early life stages of Eurasian perch *Perca fluviatilis*. Journal of Applied Ichthyology. 1996;**12**:157-165

[90] Vlavonou RS. Elevage expérimental de la perche *Perca fluviatilis* L.: Développement larvaire et croissance [thèse de doctorat]. Université de Metz, France. 1996. p. 48

[91] Vlavonou RS, Masson G, Moretteau JC. Growth of *Perca fluviatilis* larvae fed with *Artemia* spp. *Nauplii* and the effects of initial starvation. Journal of Applied Ichthyology. 1999;**15**:29-33

[92] Kestemont P, Fiogbé ED, Parfait O, Micha JC, Mélard C. Relationship between weaning size, growth, survival and cannibalism in the common perch larvae, *Perca fluviatilis*: Preliminary data. In: Lavens P, Jaseprs E, Roelants I (eds), Larvi'95, Fish and Shellfish Larviculture Symposium. Eur. Aquacult. Soc. Spe. Pub. 24, Ghent, Belgium, 1995. pp 285-288

[93] Mélard C, Baras E, Mary L, Kestemont P. Relationship between stocking density, growth, cannibalism and survival rate in intensively cultured larvae and juveniles of perch (*Perca fluviatilis*). Annales Zoologici Fennici. 1996;**33**:643-651

[94] Baras E, Kestemont P, Mélard C. Effect of stocking density on the dynamics of cannibalism in sibling larvae of *Perca fluviatilis* under controlled conditions. Aquaculture. 2003;**219**:241-255

[95] Kestemont P, Jourdan S, Houbart M, Mélard C, Paspatis M, Fontaine P, et al. Size heterogeneity, cannibalism and competition in cultured predatory fish larvae: Biotic and abiotic influences. Aquaculture. 2003;**227**:333-356

[96] Ribi G. Perch larvae (*Perca fluviatilis* L.) survived better in dilute sea water. Aquatic Sciences. 1992;**54**:85-90

[97] Bein R, Ribi G. Effects of larval density and salinity on the development of perch larvae (*Perca fluviatilis*). Aquatic Sciences. 1994;**56**:97-105

[98] Egloff M. Failure of swimbladder inflation of perch, *Perca fluviatilis* L., found in natural populations. Aquatic Sciences. 1996;**58**:15-23

[99] Mandiki RSNM, Blanchard G, Mélard C, Koskela J, Kucharczyk D, Fontaine P, et al. Effect of geographic origin on growth and food intake in Eurasian perch (*Perca fluviatilis* L.) juveniles under intensive culture conditions. Aquaculture. 2004;**229**:117-128

[100] Szczerbowski A, Luczynski MJ, Kucharczyk D. The effect of surface water spray on swimbladder inflation, survival and growth of perch (*Perca fluviatilis* L.) larvae reared under controlled conditions. In: Grizel H, Kestemont P, editors. Aquaculture and Water: Fish Culture, Shell Fish Culture and Water Usage. Aquaculture Europe'98, Bordeaux, France. European Aquaculture Society Special Publications 26. 1998. pp. 261-262

[101] Jacquemond F. Sorting Eurasian perch fingerlings (*Perca fluviatilis* L.) with and without functional swim bladder using tricaine methane sulfonate. Aquaculture. 2004;**231**:249-262

[102] Jacquemond F. Separated breeding of perch fingerlings (*Perca fluviatilis* L.) with and without initial inflated swim bladder: Comparison of swim bladder development, skeleton conformation and growth performances. Aquaculture. 2004;**239**:261-273

[103] Alix M, Zarski D, Chardard D, Fontaine P, Schaerlinger B. A detailed investigation of deformities in newly hatched embryos revealed variations among Eurasian perch populations reared in pond or RAS conditions. Journal of Zoology. 2017;**302**:126-137

[104] Alix M, Chardard D, Ledoré Y, Fontaine P, Schaerlinger B. An alternative developmental table to describe non-model fish species embryogenesis: Application to the description of the Eurasian perch (*Perca fluviatilis* L. 1758) development. EvoDevo. 2015;**6**(1):39

[105] Anthouard M, Fontaine P. L'auto-alimentation chez la perche (*Perca fluviatilis*): Adaptation à cette modalité de nourrissage et mise en évidence du rythme nycthéméral de la prise alimentaire. Ichtyophysiologica Acta. 1998;**21**:1-13

[106] Jourdan S, Fontaine P, Boujard T, Vandeloise E, Gardeur JN, Anthouard M, et al. Influence of daylength on growth, heterogeneity, gonad development, sexual steroid and thyroid levels, and N and P retention and waste in *Perca fluviatilis*. Aquaculture. 2000;**186**:253-265

[107] Mélard C, Kestemont P, Grignard JC. Intensive culture of juvenile and adult Eurasian perch (*Perca fluviatilis*): Effect of major biotic and abiotic factors on growth. Journal of Applied Ichthyology. 1996;**12**:175-180

[108] Fontaine P, Gardeur J-N, Kestemont P, Georges A. Influence of feeding level on growth, intraspecific weight variability and sexual growth dimorphism of Eurasian perch *Perca fluviatilis* L. reared in a recirculation system. Aquaculture. 1997;**157**:1-9

[109] Strand Å, Alanärå A, Staffan F, Magnhagen C. Effects of tank colour and light intensity on feed intake, growth rate and energy expenditure of juvenile Eurasian perch, *Perca fluviatilis* L. Aquaculture. 2007;**272**:312-318

[110] Strand Å, Magnhagen C, Alanärå A. Effects of repeated disturbances on feed intake, growth rates and energy expenditures of juvenile perch, *Perca fluviatilis*. Aquaculture. 2007;**265**:163-168

[111] Acerete L, Balasch JC, Espinosa E, Josa A, Tort L. Physiological responses in Eurasian perch (*Perca fluviatilis*, L.) subjected to stress by transport and handling. Aquaculture. 2004;**237**:167-178

[112] Jentoft S, Aastveit AH, Torjesen PA, Andersen Ø. Effects of stress on growth, cortisol and glucose levels in non-domesticated Eurasian perch (*Perca fluviatilis*) and domesticated

rainbow trout (*Oncorhynchus mykiss*). Comparative Biochemistry and Physiology Part A: Molecular & Integrative Physiology. 2005;**141**:353-358

[113] Juell JE, Lekang OI. The effect of feed supply rate on growth of juvenile perch (*Perca fluviatilis*). Aquaculture Research. 2001;**32**:459-464

[114] Fiogbé ED, Kestemont P. Optimum daily ration for Eurasian perch *Perca fluviatilis* L. reared at its optimum growing temperature. Aquaculture. 2003;**216**:243-252

[115] Fiogbé ED, Kestemont P, Mélard C, Micha J. The effects of dietary crude protein on growth of the Eurasian perch *Perca fluviatilis*. Aquaculture. 1996;**144**:239-249

[116] Kestemont P, Vandeloise E, Mélard C, Fontaine P, Brown P. Growth and nutritionnal status of Eurasian perch *Perca fluviatilis* fed graded levels of dietary lipids with and without added ethoxyquin. Aquaculture. 2001;**203**:85-99

[117] Xu X, Fontaine P, Mélard C, Kestemont P. Effects of dietary fat levels on growth, feed efficiency and biochemical compositions of Eurasian perch *Perca fluviatilis*. Aquaculture International. 2001;**9**:437-449

[118] Blanchard G, Makombu JG, Kestemont P. Influence of different dietary 18:3n-3/18:2n-6 ratio on growth performance, fatty acid composition and hepatic ultrastructure in Eurasian perch, *Perca fluviatilis*. Aquaculture. 2008;**284**:144-150

[119] Geay F, Kestemont P. Chapter 22. Feeding and nutrition of percid fishes during ongrowing stages. In: Kestemont P, Dabrowski K, Summerfelt RC, editors. Biology and Culture of Percid Fishes: Principles and Practices. Dordrecht, The Netherlands: Springer; 2015. pp. 587-622

[120] Lucas S. Variations de la composition chimique des tissus de la perche eurasienne élevée en circuit fermé. In: DEA Sciences Agonomiques. Nancy: INPL; 1997. 40 p

[121] Mairesse G, Thomas M, Gardeur J-N, Brun-Bellut J. Appearance and technological characteristics in wild and reared Eurasian perch, *Perca fluviatilis* (L.). Aquaculture. 2005; **246**:295-311

[122] Mairesse G, Thomas M, Gardeur J-N, Brun-Bellut J. Effects of geographic source, rearing system, and season on the nutritional quality of wild and farmed *Perca fluviatilis*. Lipids. 2006;**41**:221-229

[123] Mathis N, Feidt C, Brun-Bellut J. Influence of protein/energy ratio on carcass quality during the growing period of Eurasian perch (*Perca fluviatilis*). Aquaculture. 2003;**217**:453-464

[124] Mairesse G, Thomas M, Gardeur J-N, Brun-Bellut J. Effects of dietary factors, stocking biomass and domestication on the nutritional and technological quality of the Eurasian perch *Perca fluviatilis*. Aquaculture. 2007;**262**:86-94

[125] Gardeur JN, Mathis N, Kobilinsky A, Brun-Bellut J. Simultaneous effects of nutritional and environmental factors on growth and flesh quality of *Perca fluviatilis* using a fractional factorial design study. Aquaculture. 2007;**273**:50-63

[126] Thomas M, Mairesse G, Gardeur J-N, Brun-Bellut J. Chapter 33. Concept and determination of quality in percid fishes. In: Kestemont P, Dabrowski K, Summerfelt RC,

editors. Biology and Culture of Percid Fishes: Principles and Practices. Dordrecht, The Netherlands: Springer; 2015. pp. 843-864

[127] Rougeot C, Jacobs B, Kestemont P, Mélard C. Sex control and sex determinism study in Eurasian perch, *Perca fluviatilis*. Aquatic Living Resources. 2002;**16**:90-94

[128] Rougeot C, Nicayenzi F, Mandiki SNM, Rurangwa E, Kestemont P, Mélard C. Comparative study of the reproductive characteristics of XY males and hormonally sex-reversed XX male Eurasian perch, *Perca fluviatilis*. Theriogenology. 2004;**62**:790-800

[129] Rougeot C. Chapter 23. Sex and ploidy manipulation in percid fishes. In: Kestemont P, Dabrowski K, Summerfelt RC, editors. Biology and Culture of Percid Fishes: Principles and Practices. Dordrecht, The Netherlands: Springer; 2015. pp 625-634

[130] Rougeot C, Virumumbalu Ngingo J, Gillet J, Vanderplasschen A, Mélard C. Gynogenesis induction and sex determinism study in Eurasian perch, *Perca fluviatilis*, by use of hormonally sex-reversed male breeders. Aquaculture. 2005;**243**:411-415

[131] Rougeot C, Minet L, Prignon C, Vanderplasschen A, Detry B, Pastoret P-P, et al. Induce triploidy by heat shock in Eurasian perch, *Perca fluviatilis*, by use of hormonally sex-reversed male breeders. Aquaculture. 2003;**211**:81-89

[132] Ben Khadher S. Etude de la variabilité génétique de populations sauvages et captives de la perche eurasienne *Perca fluviatilis*, espèce en cours de domestication. Doctorat Université de Lorraine, 2015. p. 232

[133] Ben Khadher S, Agnèse J-F, Milla S, Teletchea F, Fontaine P. Patterns of genetic structure of Eurasian perch (*Perca fluviatilis* L.) in Lake Geneva at the end of the spawning season. Journal of Great Lakes Research. 2015;**41**:846-852

[134] Ben Khadher S, Fontaine P, Milla S, Agnèse J-F, Teletchea F. Genetic characterization and relatedness of wild and farmed Eurasian perch (*Perca fluviatilis*): Possible implications for aquaculture practices. Aquaculture Report. 2016;**3**:136-146

[135] Kestemont P, Mélard C. Chap. 11 - Aquaculture. In: Craig JF, editor. Percid Fishes. Systematics, Ecology and Exploitation. Oxford: Blackwell Science; 2000. pp. 191-224

[136] Kestemont P, Dabrowski K, Summerfelt RC. Biology and Culture of Percid Fishes: Principles and Practices. Vol. 901. Dordrecht, The Netherlands: Springer; 2015

[137] Fontaine P. Les marchés de la perche. Journées Techniques Black Bass, Perche, Sandre. Poisy-Chavanod, France. 1996

[138] Fontaine P. Le grossissement de la perche. Journées Techniques Black Bass, Perche, Sandre. Poisy-Chavanod, France. 1996

[139] Fontaine P. Production et marché de la perche: Bilan et perspectives. Journées Aquacoles Nationales. Nancy, France. 1997

[140] Fontaine P. Grossissement de la perche en circuit fermé. Journées Aquacoles Nationales. Nancy, France. 1997

[141] Toner D, Rougeot C. Aquaculture explained special publication—Farming of Eurasian perch. Vol. 1. Juvenile Production. Bord Lascaigh Mhara. Irish Sea Fisheries Board; 2008. 80 p

[142] Teletchea F, Fontaine P. Levels of domestication in fish: Implications for the sustainable future of aquaculture. Fish and Fisheries. 2014;**15**:181-195

[143] Bigarré L, Plassiart G, de Boisséson C, Pallandre L, Pozet F, Ledoré Y, et al. Molecular investigations of outbreaks of perch Perhabdovirus affections in pikeperch. Diseases of Aquatic Organisms. 2017;**127**:19-27

[144] Kristan J, Stejskal V, Policar T. Comparison of reproduction characteristics and broodstock mortality in farmed and wild Eurasian perch (*Perca fluviatilis* L.) females during spawning season under controlled conditions. Turkish Journal of Fisheries and Aquatic Sciences. 2012;**12**:191-197

[145] Khendek A, Alix M, Viot S, Ledoré Y, Rousseau C, Mandiki R, et al. How does a domestication process modulate oogenesis and reproduction performance in Eurasian perch? Aquaculture. 2017;**473**:206-214

[146] Douxfils J, Mandiki SNM, Marotte G, Wang N, Silvestre F, Milla S, et al. Does domestication process affect stress response in juvenile Eurasian perch *Perca fluviatilis*? Comparative Biochemistry and Physiology Part A: Molecular & Integrative Physiology. 2011;**159**:92-99

[147] Douxfils J, Mathieu C, Mandiki SNM, Milla S, Henrotte E, Wang N, et al. Physiological and proteomic evidences that domestication process differentially modulates the immune status of juvenile Eurasian perch (*Perca fluviatilis*) under chronic confinement stress. Fish and Shellfish Immunology. 2011;**31**:1113-1121

[148] Douxfils J, Desprez M, Mandiki SNM, Milla S, Henrotte E, Mathieu C, et al. Physiological and proteomic responses to single and repeated hypoxia in juvenile Eurasian perch under domestication—Clues to physiological acclimation and humoral immune modulations. Fish and Shellfish Immunology. 2012;**33**:1112-1122

[149] Douxfils J, Lambert S, Mathieu C, Milla S, Mandiki SNM, Henrotte E, et al. Influence of domestication process on immune response to repeated emersion stressors in juvenile Eurasian perch (*Perca fluviatilis* L.). Comparative Biochemistry and Physiology Part A: Molecular & Integrative Physiology. 2014;**173**:52-60

[150] Douxfils J, Mandiki SNM, Mathieu C, Milla S, Kestemont P. Chapter 29. Domestication and responses to stress. In: Kestemont P, Dabrowski K, Summerfelt RC, editors. Biology and Culture of Percid Fishes: Principles and Practices. Dordrecht, The Netherlands: Springer; 2015. pp. 743-760